ANIMAL BEHAVIOUR

BY

C. LLOYD ORGAN, F.R.S.

AUTHOR OF "ANIMAL LIFE AND INTELLIGENCE," "HABIT AND INSTINCT,"
"PSYCHOLOGY FOR TEACHERS," ETC. ETC.

ILLUSTRATED

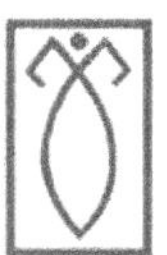

MJP PUBLISHERS

CHENNAI   NEW DELHI   TIRUNELVELI

ISBN 978-81-8094-143-6
Printed and bound in India
MJP 125
First published: 1900

MJP PUBLISHERS
New No. 5, Muthu Kalathy Street
Triplicane, Chennai 600 005
MJP's imprint: 2013

BO: 4264/ 3, First Floor, Ansari Road, Daryaganj, New Delhi-2. ©98961 92710

BO: 70B, Perumal North Car Street, Tirunelveli Junction, Tirunelveli-1 ©0462- 2330248

PREFACE

My book on "Animal Life and Intelligence" being out of print, I undertook to revise it for a new Edition. As the work of revision proceeded, however, it appeared that the amended treatment would not fall conveniently under the previous scheme of arrangement. I therefore decided to write a new book under the title of "Animal Behaviour." A few passages from the older work have been introduced, and some of the observations and conclusions already published in greater detail in "Habit and Instinct" have been summarized. But it will be found that these occupy a relatively small space in the following pages.

C. L. M.

University College, Bristol,
October 1st, 1900.

CONTENTS

CHAPTER I

ORGANIC BEHAVIOUR

		PAGE
I.	BEHAVIOUR IN GENERAL	1
II.	BEHAVIOUR OF CELLS	3
III.	CORPORATE BEHAVIOUR	14
IV.	THE BEHAVIOUR OF PLANTS	24
V.	REFLEX ACTION	31
VI.	THE EVOLUTION OF ORGANIC BEHAVIOUR	35

CHAPTER II

CONSCIOUSNESS

I.	THE CONSCIOUS ACCOMPANIMENTS OF CERTAIN ORGANIC CHANGES	42
II.	THE EARLY STAGES OF MENTAL DEVELOPMENT	48
III.	LATER PHASES IN MENTAL DEVELOPMENT	56
IV.	THE EVOLUTION OF CONSCIOUSNESS	61

CHAPTER III

INSTINCTIVE BEHAVIOUR

I.	DEFINITION OF INSTINCTIVE BEHAVIOUR	63
II.	INSTINCTIVE BEHAVIOUR IN INSECTS	71
III.	THE INSTINCTIVE BEHAVIOUR OF YOUNG BIRDS	84
IV.	THE CONSCIOUS ASPECT OF INSTINCTIVE BEHAVIOUR	98
V.	THE EVOLUTION OF INSTINCTIVE BEHAVIOUR	106

CHAPTER IV

INTELLIGENT BEHAVIOUR

PAGE

I. The Nature of Intelligent Behaviour 117
II. Intelligent Behaviour in Insects 123
III. Some Results of Experiment 134
IV. The Evolution of Intelligent Behaviour . . . 155
V. The Influence of Intelligence on Instinct . . . 168

CHAPTER V

SOCIAL BEHAVIOUR

I. Imitation 179
II. Intercommunication 193
III. Social Communities of Bees and Ants 205
IV. Animal Tradition 220
V. The Evolution of Social Behaviour 225

CHAPTER VI

THE FEELINGS AND EMOTIONS

I. Impulse, Interest, and Emotion 235
II. Play 248
III. Courtship 258
IV. Animal "Æsthetics" and "Ethics" 270
V. The Evolution of Feeling and Emotion . . . 282

CHAPTER VII

THE EVOLUTION OF ANIMAL BEHAVIOUR

I. The Physiological Aspect 295
II. The Biological Aspect 305
III. The Psychological Aspect 315
IV. Continuity in Evolution 324
Index 338

ILLUSTRATIONS

FIG.   PAGE

1. *Paramecium.* (From "Animal Biology." Longmans) . . . 4

2. Behaviour of Paramecia. (After Jennings, *American Journal of Psychology*) 8

3. Cell-division. (From " Animal Biology." Longmans) . . 13

4. Wapiti with antlers in velvet. (Drawing by Mr. Charles Whymper, after photograph by Miss Reynolds) 16

5. Wapiti with velvet shredding off. (Drawing by Mr. Charles Whymper, after photograph by Miss Reynolds) 17

6. Sun-dew leaf and tentacles. (From Darwin's "Insectivorous Plants." Murray. By kind permission of Mr. Francis Darwin, F.R.S.) 26

7. Venus's Fly-trap. (From Darwin's "Insectivorous Plants." Murray. By kind permission of Mr. Francis Darwin, F.R.S.) 27

8. Flower of *Valisneria* 28

9. Flower of *Catasetum* 30

10. Flower of *Catasetum* dissected. (From Darwin's " Fertilization of Orchids." Murray. By kind permission of Mr. Francis Darwin, F.R.S.) 31

11. Solitary Wasp stinging Caterpillar. (After Plate III. in Dr. and Mrs. Peckham's "Solitary Wasps") 75

12. Solitary Wasp dragging a Caterpillar to its Nest. (After Plate IV. in Dr. Peckham's "Solitary Wasps") 76

13. Insect Larvæ: *Sitaris, Argyromœba,* and *Leucopsis.* (After Fabre "Souvenirs") 80

14. Yucca Flower and Moth 83

FIG.   PAGE

15. Newly-hatched Chick swimming. (Drawn by Mr. Charles Whymper, after instantaneous photographs and a sketch by the author) 85

16. Nestling Megapode. (From Dr. R. Bowdler Sharpe's "Wonders of the Bird World." Wells Gardner) 87

17. Cuckoo ejecting Meadow Pipit. (From Mrs. Hugh Blackburn's sketch in "Birds from Moidart." David Douglas) . 91

18. Leaf-case of Birch-weevil 121

19. Solitary Wasp using a stone as a tool. (After Plate V. in Dr. Peckham's "Solitary Wasps") 127

20. Spiders placed by Solitary Wasps in crotches of branching stems. (After Plate X. in Dr. Peckham's "Solitary Wasps") 133

21. Fox-terrier lifting the latch of a gate. (Drawn by Mr. Charles Whymper, after a photograph by Miss Alice Worsley) . . 145

22. Cage used by Dr. Thorndike. (After figure in "Animal Intelligence," *Psychological Review*, 1898) 148

23. Diagram illustrating Dr. Thorndike's Experiments. (Based on data given in his monograph on "Animal Intelligence") 150

24. Wood ant. (From Shipley's "Invertebrates." A. & C. Black) 207

25. Beetle soliciting food from Ant. (After Wasmann. Enlarged) 213

26. Honey-pot Ant. (Enlarged) 215

ANIMAL BEHAVIOUR

CHAPTER I

ORGANIC BEHAVIOUR

I.—Behaviour in General

We commonly use the word "behaviour" with a wide range
of meaning. We speak of the behaviour of troops in the
field, of the prisoner at the bar, of a dandy in the ball-room.
But the chemist and the physicist often speak of the behaviour
of atoms and molecules, or that of a gas under changing conditions of temperature and pressure. The geologist tells us
that a glacier behaves in many respects like a river, and
discusses how the crust of the earth behaves under the stresses
to which it is subjected. Weather-wise people comment on
the behaviour of the mercury in a barometer as a storm
approaches. Instances of a similar usage need not be multi-
plied. Frequently employed with a moral significance, the
word is at least occasionally used in a wider and more comprehensive sense. When Mary, the nurse, returns with the little
Miss Smiths from Master Brown's birthday party, she is
narrowly questioned as to their behaviour ; but meanwhile
their father, the professor, has been discoursing to his students
on the behaviour of iron filings in the magnetic field ; and his
son Jack, of H.M.S. *Blunderer*, entertains his elder sisters
with a graphic description of the behaviour of a first-class
battle-ship in a heavy sea.

I

B

The word will be employed in the following pages in a wide and comprehensive sense. We shall have to consider, not only the kind of animal behaviour which implies intelligence, sometimes of a high order ; not only such behaviour as animal play and courtship, which suggests emotional attributes ; but also forms of behaviour which, if not unconscious, seem to lack conscious guidance and control. We shall deal mainly with the behaviour of the animal as a whole, but also incidentally with that of its constituent particles, or cells ; and we shall not hesitate to cite (in a parenthetic section) some episodes of plant life as examples of organic behaviour.

Thus broadly used, the term in all cases indicates and draws attention to the reaction of that which we speak of as behaving, in response to certain surrounding conditions or circumstances which evoke the behaviour. The middy would not talk of the behaviour of his ship as she lay at anchor in Portland harbour ; the word is only applicable when there is action and reaction as the vessel ploughs through a heavy sea, or when she answers to the helm. Apart from gravitation the glacier and the river would not " behave in a similar manner." Only under the conditions comprised under the term " magnetic field " do iron filings exhibit certain peculiarities of behaviour. And so, also, in other cases. The behaviour of cells is evoked under given organic or external conditions ; instinctive, intelligent, and emotional behaviour are called forth in response to those circumstances which exercise a constraining influence at the moment of action.

It is therefore necessary, in a discussion of animal behaviour, that we should endeavour to realize, as far as possible, in every case, first, the nature of the animal under consideration ; secondly, the conditions under which it is placed ; thirdly, the manner in which the response is called forth by the circumstances, and fourthly, how far the behaviour adequately meets the essential conditions of the situation.

II.—Behaviour of Cells

From what has already been said it may be inferred that our use of the term " behaviour " neither implies nor excludes the presence of consciousness. Few are prepared to contend that the iron filings in a magnetic field consciously group themselves in definite and symmetrical patterns, or that sand grains on a vibrating plate assemble along certain nodal lines because they are conscious of the effects of the bow by which the plate is set in sounding vibration. But where organic response falls under our observation, no matter how simple and direct that response may be, there is a natural tendency to suppose that the behaviour is conscious ; and where the response is less simple and more indirect, this tendency is so strengthened as to give rise to a state of mind bordering on, or actually reaching, conviction. Nor is this surprising : for, in the first place, organic responses, even the simplest, are less obviously and directly related to the interplay of surrounding circumstances ; and, in the second place, they are more obviously in relation to some purpose in the sense that they directly or indirectly contribute to the maintenance of life or the furtherance of well-being. Now where behaviour is complex and subserves an end which we can note and name, there arises the supposition that it may well be of the same nature as our own complex and conscious behaviour.

Take for example the behaviour of the Slipper-animalcule, Paramecium, one of the minute creatures known to zoologists as Protozoa. The whole animal is constituted by a single cell, somewhat less than one-hundredth of an inch in length, the form and behaviour of which may be readily studied under the microscope. Thousands may be obtained from water in which some hay has been allowed to rot. The surface of the Paramecium is covered with waving hair-like cilia, by which it is propelled through the water, while stiffer hairs may be shot out from the surface at any point where there is a local source of irritation, as indicated at the top of the accompanying figure.

Two little sacs expand and contract, and serve to drain off
water and waste products from the substance of the cell. Food
is taken in at the end of the funnel, shown in the lower part of
the figure. The cilia here work in such manner as to drive

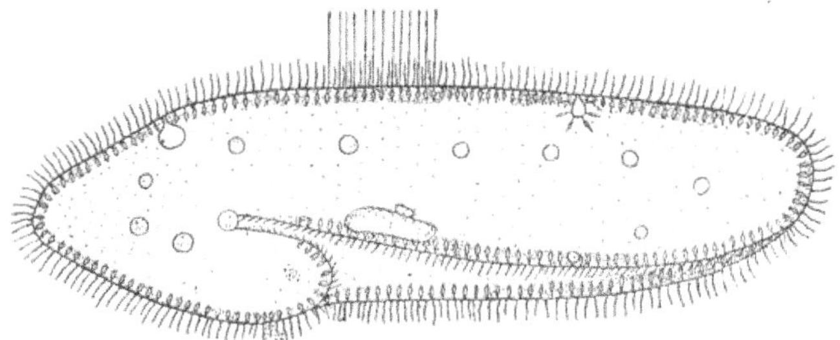

Fig. 1.—Paramecium.

the particles into and down the tube, and on reaching its inner
end these particles burst through into the semi-fluid substance, and circulate therein. Just above the funnel there are
two bean-like bodies, the larger of which is known as the
macronucleus, the smaller as the micronucleus.

The process of multiplication is by "fission," or the
division of each Paramecium into two similar animalcules.
Not infrequently, however, two Paramecia may be seen to
approach each other and come together, funnel to funnel ;
and in each the nuclei undergo curious changes. The macronucleus breaks up, and is scattered. The micronucleus in each
divides into four portions, of which three break up and dis-
appear ; while the fourth again divides into two parts, one to
be retained and the other to be exchanged for the similar
micronuclear product of the other Paramecium. The retained
portion and that received in exchange then unite to form a
new micronucleus. M. Maupas concludes from his careful
observations that, in the absence of such " conjugation " in
the mid-period of life, Paramecia pass into a state of senility
which ends in decrepitude and death. If this be so, conjugation is in them necessary for the continuance of a healthy race.

Here we have what a zoologist would describe as a specialized mode of behaviour of the nuclei ; and we have also the behaviour of the minute creatures (which contain the nuclei) as they approach each other and come together in conjugation. Can one wonder that the latter, at any rate, has been regarded as an example of conscious procedure ? In truth we do not know in what manner and by what subtle influences the Paramecia are drawn together in conjugation. But it is scarcely logical to base on such ignorance any positive assertion as to conscious attraction. It is better to confess that here is a piece of organic behaviour, the exact conditions of which are at present unexplained.

We may take from the writings * of Dr. H. S. Jennings; of Harvard, some account of other modes of behaviour among Paramecia. They largely feed upon clotted masses of bacteria. If a number are placed upon a glass slip, together with a small bacterial clot, they will be seen to congregate around the clot and to feed upon it. All apparently press in so as to reach it, or get as near it as possible. And if a number be placed on another slide without any clot, they soon collect in groups in one or more regions, as in Fig. 2, iii. It appears as if they were actuated by some social impulse leading them to crowd together and shun isolated positions. Nay, more ; it seems as if, after thus collecting and crowding in to some centre of interest, the attractive influence gradually waned ; the group spreads, and the Paramecia are less densely packed ; the assembly scatters more and more, but still seems to be retained by an invisible boundary beyond which the little creatures do not pass.

Furthermore, if kept in a jar, the Paramecia crowd up towards the surface where the bacteria clots are floating ; and if, beneath the cover glass of a slip on which they are under microscopic examination, a drop of liquid be introduced through a very fine tube, they will seem either to be attracted to it, as in Fig. 2, i., or repelled from it, as in Fig. 2, ii., according to its nature. From alkaline liquids they are

* See " The Psychology of a Protozoon," in the *Amer. Jour. of Psychology*, vol. x., No. 4, July, 1899, and the fuller papers there quoted.

repelled ; to slightly acid drops they are attracted, unless the acidity be too pungent. Heat and cold are alike repellent, and even a drop of pure distilled water forms an area into which the Paramecia do not enter.

With such facts before him, the incautious observer may be led to the conclusion that Paramecia are not only conscious, but endowed with intelligence and volition. Even M. Binet,* who occupies a position which should lead him to exercise more caution, tells us that there is not a single infusorian which cannot be frightened, and does not manifest its fear by rapid flight ; he speaks of some of these unicellular animals as "endowed with memory and volition," and possessed of "instinct of great precision ; " and he describes the following stages :—

"(1) The perception of an external object ;

"(2) The choice made between a number of objects ;

"(3) The perception of their position in space ;

"(4) Movements calculated either to approach the body and seize it, or to flee from it."

But when we seem to have grasped his point of view, when we have catalogued the memory, fear, instinct, perception, choice and volition, the whole intelligent edifice crumbles ; for we are told that "we are not in a position to determine whether these various acts are accompanied by consciousness, or whether they follow as simple physiological processes." To most of us fear, memory, choice, volition, imply something more than simple physiological processes ; they imply not only consciousness, but highly elaborated consciousness.

Dr. Jennings's researches show that no such implication can be accepted unless we are prepared to cast aside the trammels of reasonable caution. In the first place, the whole matter of feeding appears to be referable to simple organic behaviour not necessarily involving consciousness. The cilia in the mouth-groove and funnel constantly wave in such a manner as to drive a current of water, together with any

* "The Psychic Life of Micro-Organisms," 1889, p. 61.

particles which float therein, towards the interior ; and the particles are then engulphed, no matter what their composition may be. Digestible or indigestible, in they go. There is no selection of the one or rejection of the other. But, as we have seen, the Paramecia collect around a bacterial clot and feed upon it. Surely here there is selection of the nutritious ! Apparently not. They collect in just the same way towards a piece of blotting-paper, cotton-wool, cloth, sponge, or other fibrous body, and remain assembled round such an innutritious centre just as long as round a bacterial clot. There seems to be no choice in the matter : contact with any substance gives rise, as an organic response, to the lessening or cessation of the regular movements in all the cilia except those of the mouth-groove and funnel. As the Paramecia swim hither and thither, first one, then another, then more, chance to come in contact with the bacterial clot, the blotting-paper, or other substance, and since the lashing of the cilia is then automatically lessened, there they stay ; others find their way to the same spot in the course of their random movements, and they, too, stay ; thus many soon collect.

But this does not account for the seemingly social assemblages of Paramecia where there is no such substance to arrest their progress. Dr. Jennings attributes this to the fact that a dilute solution of carbon-dioxide has, what we may call for the present, an attractive influence. If a bubble of air and a bubble of carbon dioxide be introduced into the water in which Paramecia are swimming beneath a cover-glass, the animalcules collect around the carbonic dioxide, but not around the air bubble. At first they press up close to the bubble of carbon dioxide, but gradually form a ring farther and farther from its limiting boundary. This is held to be due to the fact that it is only the dilute solution of carbonic acid that has the peculiar "attraction"—a stronger solution has a different effect. And, as the gas dissolves, the Paramecia collect in a ring just where the solution is sufficiently dilute.

Now carbon dioxide is a product of the organic waste of living substance ; it is given off by active Paramecia. Where

therefore many are collected together they form a centre of
the production of this substance ; and when other Paramecia
come, in the course of their random movements, into such a

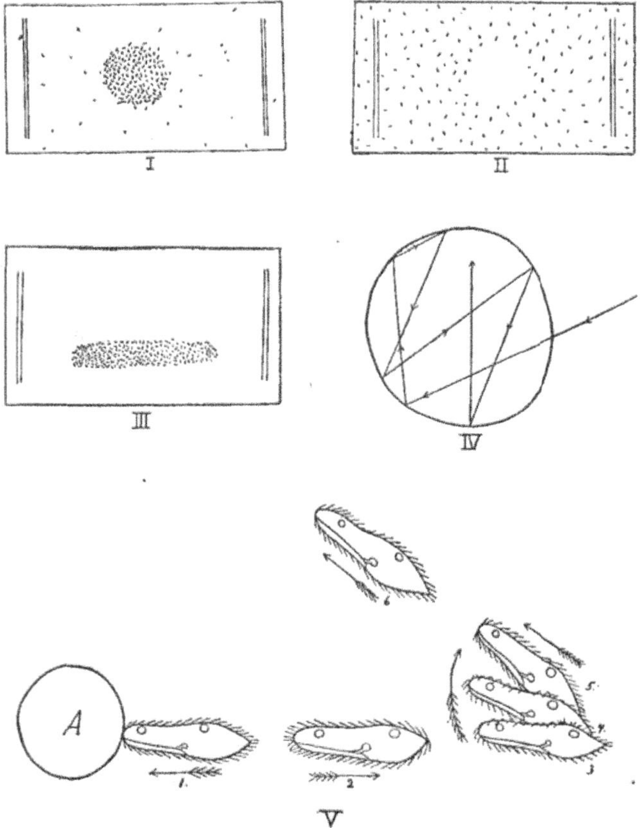

Fig. 2.—Behaviour of Paramecia (after Jennings).

centre they remain there and help to swell the numbers in the
cluster.   If Paramecia be placed in water to which a distinctly reddish tinge is given by mixing it with a small quantity

of rosol – a substance which is decolourized by carbon dioxide, and is not injurious to Paramecia—it will be seen that, where the groups are collected, the reddish tinge fades and disappears. As the groups expand, and are less densely packed, the colourless area expands too : and the limits within which the group is circumscribed are also the limits of decolourization. Dr. Jennings considers it beyond question that the assembling of Paramecia is due to the presence in such assemblages of carbonic acid produced by the animals themselves. The first beginning of the crowd may be some small fragment of bacterial clot or other substance.

It would seem, then, that Paramecia are attracted by faintly acid solutions ; and here at least there is, it may be urged, an element of choice. But even here, according to Dr. Jennings, there is not only no real choice, but not even any real attraction. What takes place, according to his observations, is briefly as follows. Suppose a faintly acid drop be inserted beneath the cover-glass. Paramecia may almost graze its boundary without being in any way affected by its presence. But in their random movements some, and eventually many, perhaps most, of the little animals chance to enter the faintly acid region ; but there is no sign of reaction or response ; they swim on across the drop until they reach its further margin. Here a reaction does take place. Instead of proceeding onwards, slowly revolving on its long axis, a Paramecium thus situated jerks backwards by a reversal of all the cilia, at the same time revolving on its axis in a direction opposite to that in which it was before turning. But the cilia of the mouthgroove resume their normal mode of working sooner than the others, and this causes the Paramecium to turn aside. It then goes ahead until it again reaches the boundary at another point, when the same behaviour is seen. The course of such a Paramecium is shown in Fig. 2, iv.

If, instead of a faintly acid drop, a little alkaline liquid be introduced beneath the cover-glass, the Paramecium similarly jerks backward and turns aside on reaching its outer boundary. The turning may carry it away from the alkali, as shown in

Fig. 2, v. ; but it just as often brings it again towards the drop, especially a large one. It seems to be a matter of chance which result follows. But eventually the little creature sails off, since each time it comes within the influence of the alkaline fluid it jerks back and turns. It appears, then, that when it is swimming in a normal solution a faintly acid liquid does not much modify its behaviour, but an alkaline fluid evokes a reversal of the cilia ; and that when it is a slightly acid solution, not only does stronger acid cause reversal, but normal fluid produces a similar result. A reaction of essentially the same kind is in fact called forth by such different stimuli as chemical substances, water heated above the normal temperature, or cooled considerably below it, and fluids which cause changes of internal pressure within the substance of the cell. Nor does it matter where the stimulus is applied. If it be applied at the hinder end the infusorian still jerks backward, though this may drive it into a destructive solution and thus cause death. There is, however, some evidence of different behaviour in some infusorians according as the stimulus is here or there. In other words, the behaviour is to some extent related to the position of the part stimulated.

Furthermore, it may be gathered from Dr. Jennings's account that there is nothing to lead us to suppose that such free living cells show any indication of what may be regarded as the keynote of intelligent behaviour. They do not profit by experience. They exhibit organic reactions which may be accompanied by some dim form of consciousness, but which do not seem to be under the guidance of such consciousness, if it exist.

One of the first lessons which the study of animal behaviour, in its organic aspect, should impress upon our minds is, that living cells may react to stimuli in a manner which we perceive to be subservient to a biological end, and yet react without conscious purpose—that is to say, automatically. The living cell assimilates food and absorbs oxygen, it grows and subdivides, it elaborates secretions, produces a skeletal framework or covering, rids itself of waste products, responds to stimuli in a definite fashion, moves hither and thither at random, its functional activities being stimulated or checked by many influences ; and yet this varied life may give no evidence of a guiding consciousness : if purpose there be, it lies deeper than its protoplasm, deeper than the dim sentience which may be present or may be absent—we cannot tell which.

And when the cells are incorporated in the body of one of the higher animals, instead of each preserving a free and nomad existence ; when they become the multitudinous constituents of an organic republic with unity of plan and unity of biological end, then the behaviour of each is limited in range but perfected within that range, in subservience to the requirements of the more complex unity. The muscle cell contracts, the gland-cell secretes, the rods and cones of the retina respond to the waves of light, and all the normal responses of the special cells go on with such orderly regularity that the term behaviour seems scarcely applicable to reactions so stereotyped. But the physiologist and the physician know well that such uniformity of response is dependent on uniformity of conditions. A little dose of some drug will profoundly modify and render abnormal the procedure which was before so mechanical in its exactitude ; and we are thus led to see how dependent the orderly behaviour really is on the maintenance of certain surrounding conditions.

Moreover, the existence of every cell in the body corporate is the outcome of a process of division involving a special mode of behaviour in the nucleus, of which we are only beginning to guess the meaning and significance, and of which we seek in vain to find an explanation in mechanical terms. And when we trace these divisions back to their primary source in the fertilized ovum, we find changes and evolutions in the nuclear matter of which it can only be said that the more they are studied the more complex and varied do they appear.

The egg, or ovum, is a single cell produced by the female, and varying much in size, according to the amount of food-yolk with which it is supplied. Like other cells, it has a nucleus, and this undergoes changes which are definitely related to the fertilization of the ovum, which we describe as the biological end. Such preparatory changes for a future contingency are especially characteristic of organic behaviour. There is nothing like it in the mineral kingdom. The nucleus divides into two parts, one of which passes out of the ovum and is lost. The nucleus again divides, and again one part passes out and is lost. Thus only one quarter of the original amount of nuclear matter remains. Now, division of the nucleus occurs whenever an animal cell divides ; but in this case (apart from details which would here be out of place) there is this difference. During the ordinary division of cells there are found in the nucleus a definite number of curved rods, and this number is constant for any given species ; but in the nucleus which remains in the ovum after three parts of its substance are lost, the number of rods has been reduced to half that which is common to the species. The egg is now ready for fertilization. A minute active cell, which is produced by the male, and which also has only half the normal number of rods, enters the ovum. The two nuclei approach each other, and give rise to the single nucleus of the fertilized ovum, which thus has the full number of rods—half of them derived from one parent, half from the other parent. The sperm cell of the male adds little to the store of protoplasm in the ovum ; but it introduces a minute body, which seems to initiate subsequent divisions of the cell. The nature of these divisions may be seen in the accompanying diagrammatic figure. In A the cell is just preparing to divide. Above the nucleus is the minute body (centrosome) just spoken of, which has already divided. In the nucleus the matter of which the rods will be constituted is net-like. In B this net-work has taken on the new form of a coiled thread, while the divided body above is associated with a spindle of delicate fibres. In C the membrane round the nucleus has disappeared, and the coiled thread has broken up into curved rods (chromosomes), four of which are shown. The two halves of the minute body form the centres of radiating stars. In D each curved rod has split along its length, and the two parts are being drawn asunder towards the centres of the two stars ; the cell itself is beginning to divide. In E the process is carried a step further, while in F the cell has completely divided into two : the rods have disappeared as such, and are replaced by a net-work ; a new nuclear membrane has been formed, and the minute body has again divided preparatory to the further division of the cell.

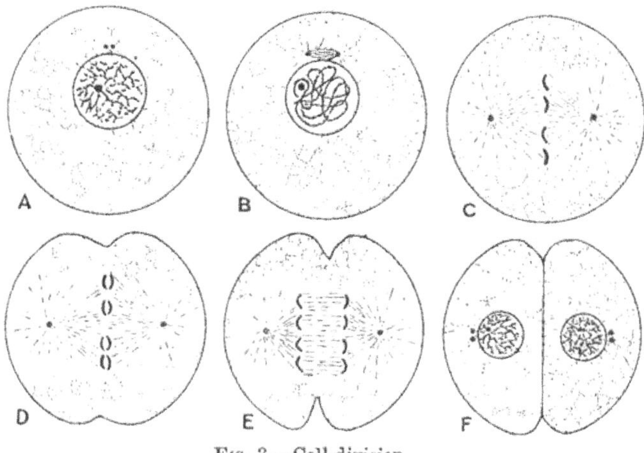

FIG. 3.—Cell-division.

Such, stripped as far as possible of technicalities, are some of the facts concerning the behaviour of cells and their nuclei during the process of cell-multiplication. No good purpose would be subserved by pretending that we fully understand them. The splitting of the rods does indeed seem an efficient means to the end of securing a fair division of the nuclear substance, which, according to many biologists, is the organic bearer of hereditary qualities in the cells. But that is nearly all that we can say. Is the process accompanied by some form of sentience ? We do not know. That it is controlled and guided by any consciousness in the cell is most improbable. But if it be a purely organic and unconscious process it should at least impress on our minds the fact that such organic be-haviour may reach a high degree of delicacy and complexity.

III. Corporate Behaviour

The word "corporate" is here applied to the organic behaviour of cells when they are not independent and free, but are incorporated in the animal body, and act in relation to each other. If the behaviour of the individual cell during division impresses us with the subtle intricacy of organic processes, the behaviour of the growing cell-republic during the early stages of organic development must impress us no less forcibly. We place the fertilized egg of a hen in an incubator, and supply the requisite conditions of warmth, moisture, and fresh air. Before the egg is laid cell-division has begun. A small patch of closely similar cells has formed on the surface of the yolk. Further subdivision is then arrested until the warmth of incubation quickens again the patch into life. But when once thus quickened no subsequent temporary arrest is possible—life will not again lie dormant. If arrest there be it is that of death. And from that little patch of cells, which spreads further and further over the yolk, a chick is developed. Into the intricate technicalities of embryology this is not the place to enter. But it is a matter of common knowledge that, whereas we have to-day an egg such as we eat for breakfast, three weeks hence we shall have a bright active bird, a cunningly wrought piece of mechanism, and, more than that, a going machine. During this wonderful process the cellular constituents take on new forms and perform new functions, all in relationship to each other, all as part of one organic whole. Here bones are developed to form a skeletal framework, there muscles are constituted which shall render orderly movements possible ; feathers, beak, and claws take shape as products of the skin ; gut and glands prepare for future modes of nutrition ; heart and blood-vessels undergo many changes, some reminiscent of bygone and ancestral gill-respiration, some in relation to the provisional respiration of the embryo by means of a temporary organ that spreads out beneath the shell, some preparatory to the future use of the lungs,—some, again, related to the absorption of food from the yolk, others to subsequent means of digestion ; nerve, brain, and sense-organs differentiate. A going machine in the egg, the chick is hatched, and forthwith enters on a wider field of behaviour. Few would think of attributing to the consciousness of the embryo chick any guiding influence on the development of its bodily structure, any control over the subtle changes and dispositions of its constituent cells. But no sooner does the chick, when it is hatched, begin to show wider modes of instinctive behaviour, than we invoke conscious intelligence for their explanation, seemingly forgetful of the fact that there is no logical ground for affirming that, while the marvellous delicacies of structure are of unconscious organic origin, the early modes of instinctive behaviour are due to the guidance of consciousness. Such modes of behaviour will, however, be considered in another chapter. Here we have to notice that the unquestionably organic behaviour of the incorporated republic of cells may attain to a high degree of complexity, and may serve a distinctly biological end.

There is, perhaps, no more striking instance of rapid and vigorous growth than is afforded by the antlers of deer,* which are shed and renewed every year. In the early summer, when growing, they are covered over with a dark hairy skin, and are said to be " in velvet." If you lay your hand on the growing antler, you will feel that it is hot with the nutrient blood that is coursing beneath it. It is, too, exceedingly sensitive and tender. An army of tens of thousands of busy living cells is at work beneath that velvet surface, building the bony antlers, preparing for the battles of autumn. Each minute cell, working for the general good, takes up from the nutrient blood the special materials it requires ; elaborates the crude bone-stuff, at first soft as wax, but ere long to become hard as stone ; and then, having done its work, having added its special morsel to the fabric of the antler, remains embedded and immured, buried beneath the bone-products of its successors or descendants. No hive of bees is busier or more replete with active life than

---

* This paragraph is taken from " Animal Life and Intelligence," p. 28.

the antler of a stag as it grows beneath the soft, warm velvet.
And thus are built up in the course of a few weeks those
splendid " beams," with their "tynes " and " snags," which, in
the case of the wapiti, even in the confinement of our Zoological

FIG. 4.—Wapiti with antlers in velvet.

Gardens, may reach a weight of thirty-two pounds, and which,
in the freedom of the Rocky Mountains, may reach such a size
that a man may walk, without stooping, beneath the archway
made by setting up upon their points the shed antlers.   When
the antler has reached its full size, a circular ridge makes its

appearance at a short distance from the base. This is the
" burr," which divides the antler into a short " pedicel " next
the skull, and the " beam " with its branches above. The
circulation in the blood-vessels of the beam now begins to
languish, and the velvet dies and peels off, leaving the hard,

Fig. 5.—Wapiti with velvet shredding off.

bony substance exposed. Then is the time for fighting, when
the stags challenge each other to single combat, while the
hinds stand timidly by. But when the period of battle is over,
and the wars and loves of the year are past, the bone beneath
the burr begins to be eaten away, through the activity of
certain large bone-absorbing cells, and, the base of attachment

being thus weakened, the antlers are shed ; the scarred surface skins over and heals, and only the hair-covered pedicel of the antler is left.

We have no reason to suppose that this corporate cellular behaviour, involving the nicely adjusted co-operation of so vast an army of organic units, is under the conscious guidance of the stag. And yet how orderly the procedure ! how admirable the result ! Nor is there an organ or structural part of the stag or any other animal that does not tell the same tale. This is but one paragraph of the volume in which is inscribed the varied and wonderful history of organic behaviour in its corporate aspect. Is it a matter for wonder that the cause of such phenomena has been regarded as "a mystery transcending naturalistic conception ; as an alien influx into nature, baffling scientific interpretation " ? And yet, though not surprising, this attitude of mind, in face of organic phenomena, is illogical, and is due partly to a misconception of the function of scientific interpretation, partly to influences arising from the course pursued by the historical development of scientific knowledge. The function of biological science is to formulate and to express in generalized terms the related antecedences and sequences which are observed to occur in animals and plants. This can already be done with some approach to precision. But the underlying cause of the observed phenomena does not fall within the purview of natural science ; it involves meta-physical conceptions. It is no more (and no less) a "mystery" than all causation in its last resort—as the *raison d'être* of observed phenomena—is a mystery. Gravitation, chemical affinity, crystalline force,—these are all " mysteries."

If the mystery of life, lying beneath and behind organic behaviour, be said to baffle scientific interpretation, this is because it suggests ultimate problems with which science as such should not attempt to deal. The final causes of vital phenomena (as of other phenomena) lie deeper than the probe of science can reach. But why is this sense of mystery especially evoked in some minds by the contemplation of organic behaviour, by the study of life ? Partly, no doubt, because the scientific interpretation of organic processes is but recent, and in many respects incomplete. People have grown so accustomed to the metaphysical assumptions employed by physicists and chemists when they speak of the play of crystalline forces and the selective affinities of atoms, they have been wont for so long to accept the "mysteries" of crystallization and of chemical union, that these assumptions have coalesced with the descriptions and explanations of science ; and the joint products are now, through custom, cheerfully accepted as natural. Where the phenomena of organic behaviour are in question, this coalescence has not yet taken place ; the metaphysical element is on the one hand proclaimed as inexplicable by natural science, and on the other hand denied even by those who talk glibly of physical forces as the final cause of the phenomena of the inorganic world.

So much reference to the problems which underlie the problems of science seems necessary. It is here assumed that the phenomena of organic behaviour are susceptible of scientific discussion and elucidation. But even assuming that an adequate explanation in terms of antecedence and sequence shall be thus attained by the science of the future, this will not then satisfy, any more than our inadequate explanations now satisfy, those who seek to know the ultimate meaning and reason of it all: What makes organic matter behave as we see it behave ? what drives the wheels of life, as it drives the planets in their courses ? what impels the egg to go through its series of developmental changes ? what guides the cells along the divergent course of their life-history ? These are questions the ultimate answers to which lie beyond the sphere of science —questions which man (who is a metaphysical being) always does and always will ask, even if he rests content with the answer of agnosticism ; but questions to which natural science never will be able, and should never so much as attempt, to give an answer.

Enough has now been said to show that organic behaviour is a thing *sui generis*, carrying its own peculiar marks of distinction ; and further, that, for science, this is just part of the constitution of nature, neither more nor less mysterious than, let us say, crystallization or chemical combination. But associated and closely interwoven with all that is distinctively organic there is much which can to some extent be interpreted in terms of physics and chemistry.

The animal * has sometimes been likened to a steam-engine, in which the food is the fuel which enters into combustion with the oxygen taken in through the lungs. It may be worth while to modify and modernize this analogy—always remembering, however, that such an analogy must not be pushed too far.

In the ordinary steam-engine the fuel is placed in the fire-box, to which the oxygen of the air gains access; the heat produced by the combustion converts the water in the boiler into steam, which is made to act upon the piston, and thus set the machinery in motion. But there is another kind of engine, now extensively used, which works on a different principle. In the gas-engine the fuel is gaseous, and it can thus be introduced in a state of intimate mixture with the oxygen with which it is to unite in combustion. This is a great advantage. The two can unite rapidly and explosively. In gunpowder the same end is effected by mixing the carbon and sulphur with nitre, which contains the oxygen necessary for their explosive combustion. And this is carried still further in dynamite and gun-cotton, where the elements necessary for explosive combustion are not merely mechanically mixed, but are chemically combined in a highly unstable compound.

But in the gas-engine, not only are the fuel and the oxygen thus intimately mixed, but the controlled explosions are caused to act directly on the piston, and not through the intervention of water in a boiler. Whereas, therefore, in the steam-engine the combustion is to some extent external to the working of the machine, in the gas-engine it is to a large extent internal and direct.

Now, instead of likening the animal as a whole to a steam-engine, it is more satisfactory to liken each cell to an automatic

* The following paragraphs are taken with some slight changes from " Animal Life and Intelligence," pp. 30–35.

gas-engine which manufactures its own explosive. During
the period of repose which intervenes between periods of
activity, its protoplasm is busy in construction, taking from
the blood-discs oxygen, and from the blood-fluid carbonaceous
and nitrogenous materials, and knitting these together into
relatively unstable explosive compounds, which play the part
of the mixed air and gas of the gas-engine. A resting muscle
may be likened to a complex and well-organized battery of
gas-engines. On the stimulus supplied through a nerve-
channel a series of co-ordinated explosions takes place : the
gas-engines are set to work ; the muscular fibres contract ;
the products of the silent explosions are taken up and hurried
away by the blood-stream ; and the protoplasm prepares a
fresh supply of explosive material. Long before the invention
of the gas-engine, long before gun-cotton or dynamite were
dreamt of, long before some Chinese or other inventor first
mixed the ingredients of gunpowder, organic nature had utilized
the principle of controlled explosions in the protoplasmic cell,
and thus rendered animal behaviour possible.

Certain cells are, however, more delicately explosive than
others. Those, for example, on or near the external surface
of the body—those, that is to say, which constitute the end-
organs of the special senses—contain explosive material which
may be fired by a touch, a sound, an odour, the contact with
a sapid fluid or a ray of light. The effects of the explosions
in these delicate cells, reinforced in certain neighbouring nerve-
batteries, are transmitted down the nerves as waves of subtle
chemical or electrolytic change, and thus reach that wonderful
aggregation of organized and co-ordinated explosive cells, the
brain. Here it is again reinforced and directed (who, at
present, can say how ?) along fresh nerve-channels to muscles,
or glands, or other organized groups of explosives. And in
the brain, somehow associated with the explosion of its cells,
consciousness, the mind-element, emerges ; of which we need
only notice here that it belongs to a *wholly different order of
being* from the physical activities and products with which we
are at present concerned.

We must not press the explosion analogy too far. The essential thing seems to be that the protoplasm of the cell has the power of building up complex and unstable chemical compounds, which are perhaps stored in its spongy substance ; and that these unstable compounds, under the influence of a stimulus (or, possibly, sometimes spontaneously), break down into simpler and more stable compounds. In the case of muscle-cells, this latter change is accompanied by an alteration in length of the fibres, and consequent movements in the animal, the products of the disruptive change being useless or harmful, and being, therefore, removed as soon as possible. But very frequently the products of explosive activity are made use of. In the case of bone-cells, one of the products of disruption is of permanent use to the organism, and constitutes the solid framework of the skeleton. In the case of the secreting cells in the salivary and other digestive glands, some of the disruptive products are of temporary value for the preparation of the food. It is probable that these useful products of disruption, permanent or temporary, took their origin in waste products for which natural selection has found a use, and which have been gradually rendered more and more efficacious in modes of organic behaviour increasingly complex.

In the busy hive of cells which constitutes what we call the animal body, there is thus ceaseless activity. During periods of apparent rest the protoplasm is engaged in constructive work, building up fresh supplies of unstable materials, which, during periods of apparent activity, break up into simpler and more stable substances, some of which are useful to the organism, while others must be got rid of as soon as possible. From another point of view, the cells during apparent rest are storing up energy to be utilized by the animal during its periods of activity. The storing up of available energy may be likened to the winding up of a watch or clock ; it is when an organ is at rest that the cells are winding themselves up ; and thus we have the apparent paradox that the cell is most active and doing most work when the organ of which it forms a part is at rest. During the repose of an organ, in fact, the cells are busily working in preparation for the manifestation of energetic action that is to follow. Just as the brilliant display of intellectual activity in a great orator is the result of the silent work of a lifetime, so is the physical manifestation of muscular power the result of the silent preparatory work of the muscle-cells.

It may, perhaps, seem strange that the products of cellular life should be reached by the roundabout process of first producing unstable compounds, from which are then formed more stable substances, useful for permanent purposes as in bone, or temporary purposes as in the digestive fluids. It seems a waste of power to build up substances unnecessarily complex and stored with an unnecessarily abundant supply of energy. But only thus could the organs be enabled to act under the influence of stimuli, and afford examples of corporate behaviour. They are like charged batteries ready to discharge under the influence of the slightest organic touch. In this way, too, is afforded a means by which the organ is not dependent only upon the products of the immediate activity of the protoplasm at the time of action, but can utilize the store laid up during preceding periods of rest.

Sufficient has now been said to illustrate the nature of some of the physical processes which accompany organic behaviour in its corporate aspect. The fact that should stand out clearly is that the animal body is stored with large quantities of available energy resident in highly complex and unstable chemical compounds, elaborated by the constituent cells. These unstable compounds, eminently explosive according to our analogy, are built up of materials derived from two different sources—from the nutritive matter (containing carbon, hydrogen, and nitrogen) absorbed during digestion, and from oxygen taken up from the air during respiration. The cells thus become charged with energy that can be set free on the application of the appropriate stimulus, which may be likened to the spark that fires the explosive.

Let us note, in conclusion, that it is through the blood-system, ramifying to all parts of the body, and the nerve-system, the ramifications of which are not less perfect, that one of the larger and higher animals is knit together into an organic whole. The former carries to the cell the raw materials for the elaboration of its explosive products, and, after the explosions, carries off the waste products which result therefrom. The nerve-fibres carry the stimuli by which the explosive is fired, while the central nervous system organizes, co-ordinates, and controls the explosions, and initiates the elaboration of the explosive compounds. Blood and nerves co-operate to render corporate behaviour possible.

IV.—The Behaviour of Plants

A short parenthetic section on the behaviour of plants may serve further to illustrate the nature of organic behaviour. We have seen that Paramecium is apparently attracted by faintly acid solutions, and have briefly considered Dr. Jennings's interpretation of the facts disclosed by careful observation. In the ferns the female element, or ovum, is contained in a minute flask-shaped structure (archegonium), in the neck and mouth of which mucilaginous matter, with a slightly acid reaction, is developed ; and this is said to exercise an attractive influence on the freely swimming ciliated male elements, or spermatozoids, which are necessary for fertilization. " Now, it has been shown by experiment that the spermatozoids of ferns are attracted by certain chemical substances, and especially by malic acid. If artificial archegonia are prepared (consisting of tiny capillary glass-tubes) and filled with mucilage to which a small quantity of this acid has been added, they are found, when placed in water containing fern-spermatozoids, to exercise the same attraction upon them which the real archegonia exercise in nature. The malic acid gradually diffuses out into the water, and the spermatozoids are influenced by it, so that they move in the direction in which the substance is more concentrated, i.e. towards the tube. Although it cannot be proved that the archegonia themselves contain malic acid, as they are too small for a recognizable quantity to be obtained

from them, yet there can be little doubt that the natural
archegonia owe their attractive influence to the same chemical
agent which has proved efficacious in experiment."* In the
light of Dr. Jennings's observations, it is perhaps not improbable that this so-called attractive influence is similar to
that seen in Paramecium ; and that the spermatozoids enter
the organic acid in the course of their random movements, and
there remain. Be that as it may, the male elements collect in
the mucilaginous mass, and pass down the neck of the flask
until one reaches and coalesces with the female element, or
ovum, and effects its fertilization. Here we have organic
behaviour unmistakably directed to a biological end—behaviour
which may indeed be accompanied by some dim form of
consciousness, but which is due to a purely organic reaction.
It is scarcely satisfactory to say that the spermatozoids " possess
a certain power of perception, by which their movements are
guided." † If consciousness be present, it is probably merely
an accompaniment of the response, and has no directive influence on its nature and character.

In the higher plants, as in the higher animals, the
differentiation and the orderly marshalling of the cell-progeny
arising from the coalescent male and female elements, afford,
during development, examples of corporate organic behaviour
which can be more readily described than explained, but which
not less clearly subserve definite biological ends, and in many
cases, such as the direction of growth in radicles and roots, the
curling of tendrils, and the reaction to the influence of light
and warmth, are related to and evoked by the environing conditions. More closely resembling familiar modes of behaviour
in animals are such movements as are seen in the " tentacles "
which project from the upper surface and margin of the Sun-
dew leaf. Their knobbed ends secrete a sticky matter, which
glistens in the sun, and to which small foreign bodies readily
adhere. If particles of limestone, sand, or clay, such as may

* D. H. Scott, " An Introduction to Structural Botany," part ii.,
" Flowerless Plants," pp. 70, 71.

† *Ibid.*, p. 71.

be blown by the wind, touch and stick to these knobs, there follows an exudation of acid liquid, but no marked and continuous change occurs in the position of the tentacles. But should an insect alight on the leaf, or a small piece of meat be placed upon the tentacles, not only is there a discharge of acid juice, but a ferment is also produced, which has a digestive action on the nitrogenous matter. Slowly the tentacle curves inwards and downwards, as one's finger may bend towards the palm of one's hand; neighbouring tentacles also turn towards and incline on to the stimulating substance; then others, further off, behave in a similar way, until all the tentacles, some two hundred in number, are inflected and converge upon the nitrogenous particle. Nay, more : "When two little bits of meat are placed simultaneously on the right and left halves of the same Sun-dew leaf, the two hundred tentacles divide into two groups, and each one of the groups directs its aim to one of the bits of meat." *

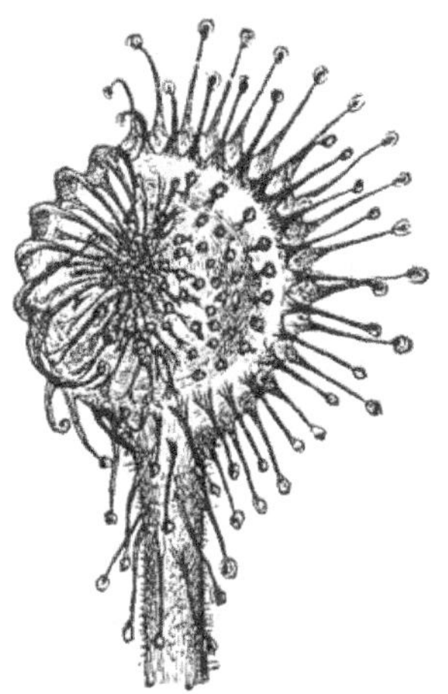

Fig. 6.— Sun-dew (*Drosera*). Leaf (enlarged) with the tentacles on one side inflected over a bit of meat placed on the disc. (From Darwin's " Insectivorous Plants.")

The movements, though slow, are orderly, methodical, and effective, the secretions of many glands being brought to bear on just those substances which are capable of digestion and absorption by the plant. The seemingly concerted action is moreover due to an organic transmission of impulses from cell to cell—a transmission accompanied by visible changes in a purple substance contained within the cells. In the Sun-dew any tentacle may form the starting-point of the spreading wave of impulse. But in the Venus's Fly-trap there are six

* Kerner, " Natural History of Plants," translated by F. W. Oliver, vol. i., p. 145.

delicate spines, the slightest touch on any one of which causes the two halves of the specially modified leaf-end to fold inwards on the midrib as a hinge. The transmission of impulse is more rapid, the trap closing in a few seconds ; and electric currents have been observed to accompany the change. Tooth-like spines at the edge of the trap interlock, and serve to prevent the escape of small insects, while short-stalked purple glands secrete an acid digestive juice. Division of labour has been carried further ; and organic behaviour, not less pur-posive, is carried out in a manner even more effective.

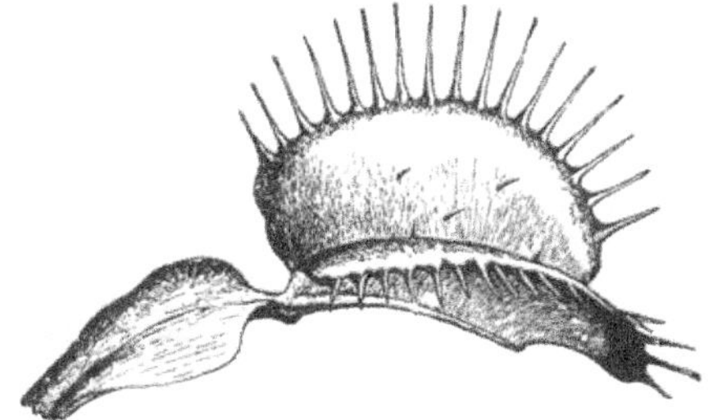

Fig. 7.—Venus's Fly-trap (*Dionæa*). Leaf viewed laterally in its ex-panded state. (From Darwin's " Insectivorous Plants.")

In other plants adaptive movements are well known. " Few phenomena have such a peculiar appearance as the movements which occur in the sensitive Oxalis when rain comes on. Not only do the leaflets on which the finest rain-drops fall fold together in a downward direction, but all the neigh-bouring ones perform the same movement, although they have not themselves been shaken by the impact of the falling drops. The movement is continued to the common leaf-stalk bearing the numerous leaflets. This also bends down towards the ground. The rain-drops now slide over the bent leaf-stalk and down over the depressed leaflets, and not a drop remains behind on their delicate surfaces." * The waves of impulse are said to be transmitted along definite lines, and to cause

* Kerner, " Natural History of Plants," vol. i., p. 536.

the expulsion of water from certain cells at the point of insertion of the leaflets or leaf-stalks, rendering them flaccid.

Appealing even more strongly to the popular imagination, though probably not of deeper biological significance, is the behaviour of plants in relation to the essential process of fertilization. Only two examples can here be cited. *Valisneria spiralis* is an aquatic plant, with long submerged strap-like leaves, which grows in still water in Southern Europe. The female flower is enclosed in two translucent bracts, which form a protective bladder so long as the flower is beneath

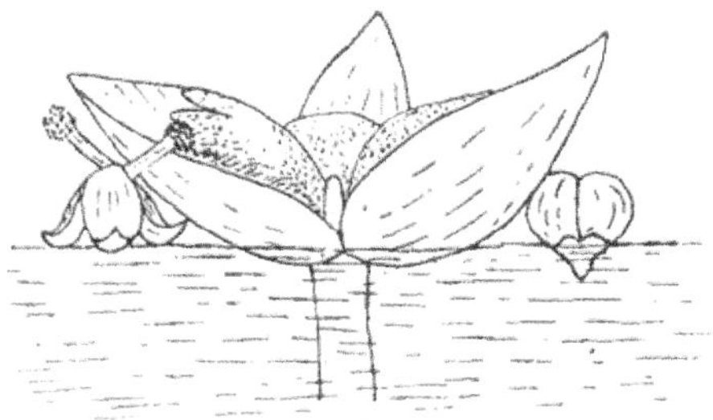

Fig. 8.—Flower of *Valisneria*.

the surface of the water ; but the flower-stalk continues to grow until the flower reaches the surface, when it becomes freely exposed by the splitting of the bracts. There are three boat-shaped sepals, which act as floats ; three quite minute petals ; and three large fringed stigmas, which project over the abortive petals in the space between the boat-like sepals. The flower is now ready for fertilization.

The male flowers, which are developed on different individuals from those which produce the female flowers, grow in bunches beneath an investing bladder. The stalk does not elongate, so that the bladder never rises far above the bottom, and remains completely submerged. Here the bladder bursts, and the male flowers, with short stalks, are detached. Each has three sepals, which enclose and protect the stamens. The separated flower now ascends to the surface, the sepals open and form three hollow boats, by means of which the flower floats freely, while the two functional stamens project upwards and somewhat obliquely into the air, exposing the large sticky pollen-cells. Blown hither and thither by the wind, these little flower-boats "accumulate in the neighbourhood of fixed bodies, especially in their recesses, where they rest like ships in harbour. When the little craft happen to get stranded in the recesses of a female Valisneria flower, they adhere to the tri-lobed stigma, and some of the pollen-cells are sure to be left sticking to the fringes on the margins of the stigmatic surface." *

This is a good example of purely organic behaviour admirably adapted to secure a definite and important biological end. Few will be likely to contend that it is even accompanied by, still less under the guidance of, any conscious foresight on the part of the plant. And the lesson it should teach is that, in the study of organic behaviour, adaptation to the conditions of existence is not necessarily the outcome of conscious guidance.

It is well known that the orchids exhibit, in their mode of fertilization, remarkable adaptations by which the visits of insects are rendered subservient to the needs of the plant. In the Catasetums, for example, the male flower may be described as consisting of two parts—a lower part, the cup-like labellum (Figs. 9, *l*), which constitutes a landing-stage on which insects may alight ; and an upper part, the column (Fig. 9, *c*), surrounded by the upper sepal and petals. In the upper part of the column the pollen-masses are borne at one end of an elastic pedicel, at the other end of which is an adhesive disc, and the rod is bent over a pad so as to be in a state of strain. The disc is retained in position by a membrane with which two long tubular horns (Figs. 9, *h* ; 10, *an*) are continuous. These project over the labellum, where insects alight to gnaw its sweet fleshy walls, and if they be touched, even very lightly, they convey some stimulus to the membrane which surrounds and connects the disc with the adjoining surface,

* Kerner, "Natural History of Plants," vol. ii., p. 132.

causing it instantly to rupture ; and as soon as this happens, the disc is suddenly set free. The highly elastic pedicel then flirts the disc out of its chamber with such force that the whole is ejected, sometimes to a distance of two or three feet,

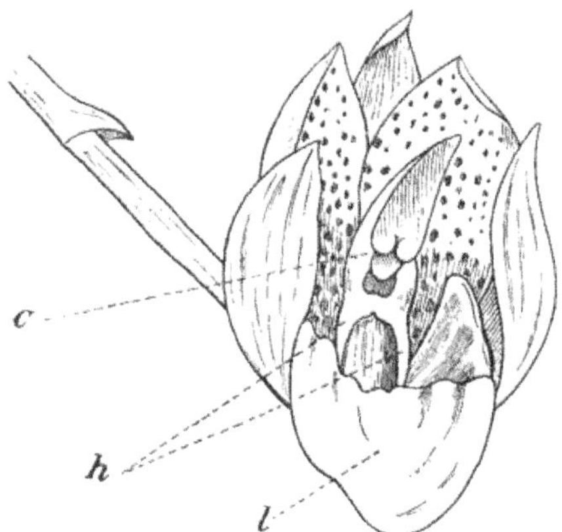

Fig. 9.—Flower of *Catasetum* ; c, column; h, horns; l, labellum.

bringing away with it the two pollen-masses. " The utility of so forcible an ejection is to drive the soft and viscid cushion of the disc against the hairy thorax of the large hymenopterous insects which frequent the flowers. When once attached to an insect, assuredly no force which the insect could exert would remove the disc and pedicel, but the caudicles [by which the pollen-masses are attached] are ruptured without much difficulty, and thus the balls of pollen might readily be left on the adhesive stigma of the female flower." *

Here again we have adaptive behaviour of exquisite nicety, and we have the transmission of an impulse very rapidly along the cells of the irritable horns, followed by the sudden rupture of a membrane. Beautiful, however, as is the adaptation, effective as it is to a definite biological end, the organic behaviour does not afford any indication of the guidance of

* Darwin, " Fertilization of Orchids," 2nd edit., pp. 191, 192.

consciousness. Among plants we have many interesting and admirable examples of organic behaviour; but nowhere so much as a hint of that profiting by individual experience

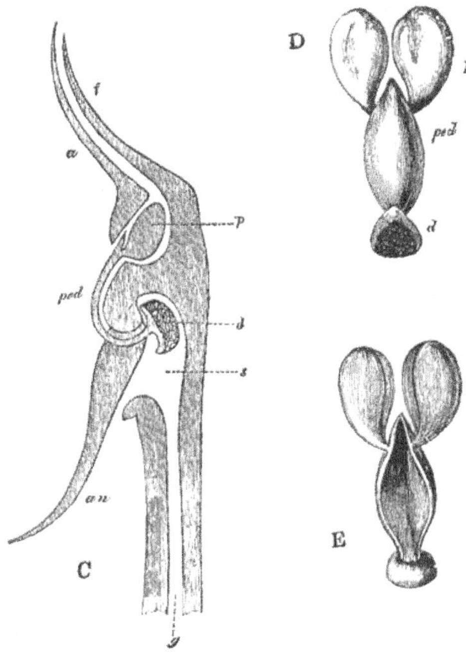

Fig. 10.—*Catasetum*; C, diagram of column; *a*, anther; *an*, horn; *d*, adhesive disc; *f*, filament of anther; *g*, ovarium; *ped*, pedicel; D and E, pollinium; *p*, pollen-mass. (From Darwin's " Orchids.")

which is the criterion of the effective presence of conscious guidance and control.

V.—Reflex Action

It is sometimes said that the tentacles of the Sun-dew leaf indicate a primitive kind of reflex action in plants, and that they afford evidence of discrimination. " It is," says Romanes, "the stimulus supplied by continuous *pressure* that is so delicately perceived, while the stimulus supplied by *impact* is disregarded." * And, comparing this with what is observed in the Venus's Fly Trap, he says : "In these two plants the power of discriminating between these two kinds of stimuli has been developed to an equally astonishing extent, but in opposite directions." † It is well, however, to avoid terms which carry with them so distinctively a conscious implication as "discrimination" and "perception" do for most of us. Just as the photographer's film reacts differently according to the quality of light-rays, violet or red, which reach it, so do many organic substances react differently to stimuli of different quality, irrespective of their intensity. The "discrimination" of plants and of some of the lower animals is of this kind, and it is better to speak of it simply as differential reaction. There can then be no chance of its being confused with conscious choice.

Nor should the movements of the Sun-dew tentacles or of those of the Sea-anemone be termed in strictness reflex action. As originally employed by Marshall Hall, and, since that time, by common consent, *reflex action* involves a differentiated nervous system. There is, first, an afferent impulse from the point of stimulation passing inwards to a nerve-centre ; secondly, certain little-understood changes within this centre ; and thirdly, an efferent impulse from the centre to some organ or group of cells which are thus affected. In plants there is no indication of anything analogous to this specialized mode of response. The impulse passes directly from the point of stimulation to the part affected without the intervention of anything like a nerve-centre. In the sensitive Oxalis the impulse passes directly to the point of insertion of the leaflet or leaf-stalk ; in Catasetum, from the horn to the retaining membrane ; in the Sun-dew, from the affected tentacle to those in its neighbourhood. Even in the Sea-anemone, though there is a loosely diffused nervous system, the passage

* "Mental Evolution in Animals," p. 50.
† *Ibid.*, p. 51.

of the impulse from one part of the circlet of tentacles to
other parts, seems to follow a direct rather than a reflex
course, and there do not appear to be any specialized centres
by which the impulses are received and then redistributed.

In all animals in which well-differentiated nervous systems
are found, in which there are distinct nerve-fibres and nerve-
centres, reflex actions, simple or more complicated, occur.
They form the initial steps leading up to the highest types
of organic behaviour. So long as the nervous arcs—afferent
fibres, nerve-centre, and efferent fibres—remain intact reflex
acts may be carried out with great precision and delicacy,
even when the higher centres, which we believe to be those of
conscious guidance and control, have been destroyed. When,
for example, the whole of the brain of a frog has been extir-
pated and the animal is hung up by the lower jaw, if the left
side be touched with a drop of acid the left leg is drawn up
and begins to scratch at the irritated spot, and when this leg
is held, the other hind leg is, with seemingly greater difficulty,
brought to bear on the same spot. "This," says Sir Michael
Foster, "at first sight looks like an intelligent choice. . . .
But a frog deprived of its brain so that the spinal cord only
is left, makes no spontaneous movements at all. Such an
entire absence of spontaneity is wholly inconsistent with the
possession of intelligence. . . . We are therefore led to conclude that the phenomena must be explained in some other
way than by being referred to the working of an intelligence." * But if we concede that intelligence is absent, may
there not at least be some consciousness? Sir Michael
Foster's reply to such a question goes as far as we have any
justification for going, even when we give free rein to conjecture. "We may distinguish," he says, "between an active
continuous consciousness, such as we usually understand by
the term, and a passing or momentary condition, which we
may speak of as consciousness, but which is wholly discon-
tinuous from an antecedent or from a subsequent similar
momentary condition; and indeed we may suppose that the

* "A Text-book of Physiology," 5th edit., part iii., p. 909.

complete consciousness of ourselves, and the similarly complete consciousness which we infer to exist in many animals, has been evolved out of such a rudimentary consciousness. We may, on this view, suppose that every nervous action of a certain intensity or character is accompanied by some amount of consciousness which we may, in a way, compare to the light emitted when a combustion previously giving rise to invisible heat waxes fiercer. We may thus infer that when the brainless frog is stirred by some stimulus to a reflex act, the spinal cord is lit up by a momentary flash of consciousness coming out of the darkness and dying away into darkness again ; and we may perhaps infer that such a passing consciousness is the better developed the larger the portion of the cord involved in the reflex act and the more complex the movement. But such a momentary flash, even if we admit its existence, is something very different from consciousness as ordinarily understood, is far removed from intelligence, and cannot be appealed to as explaining the 'choice' spoken of above." *

These sentences indicate with sufficient clearness the distinction, more than once hinted at in the foregoing pages, between consciousness as an accompaniment, and consciousness as a guiding influence. We shall have more to say in this connection in subsequent chapters. The experiment with the frog shows, at any rate, that reflex actions, of a distinctly purposive nature, may be carried out when the centres, which are believed to exercise conscious control and guidance have been destroyed. It is said that in man, when, owing to injuries of the spine, the connection between the brain and the lower part of the spinal cord have been severed, tickling of the foot causes withdrawal of the limb without directly affecting the consciousness of the patient. But in all such cases we are dealing with a maimed creature. The living frog or man, healthy and intact, is, presumably in the one case, certainly in the other, conscious of these reflex actions, and can exercise some amount of guidance and control over them. In man this is unquestionably the case. But granting that

* " A Text-book on Physiology," 5th edit., part iii., pp. 911, 912.

he brain is the organ of conscious control, granting that it
an receive impulses from and transmit impulses to the reflex
entres, no more is here implied, and no more can be legiti-
nately inferred, than that the kind of organic behaviour we
all " reflex action " is in the higher animals in touch with
he guiding centres. We have no ground for assuming that
n reflex action there is any power of intelligent guidance
ndependent of that which is exercised by the brain or
analogous organ. In brief, reflex acts, in animals endowed
vith intelligence, may be regarded as specialized modes of
organic behaviour; which are in themselves often charac-
erized by much complexity; which subserve definite biological
ends; which are effected by subordinate centres capable of
transmitting impulses to, and receiving impulses from, the
entres of intelligent guidance; and which, as responses con-
ined to certain organs or parts of the body, form elements
in the wider behaviour of the animal as a whole.

VI.—THE EVOLUTION OF ORGANIC BEHAVIOUR

The interpretation of organic behaviour in terms of evolu-
ion mainly depends on the answer we give to the question :
Are acquired modes of behaviour inherited? A negative
answer to this question is here provisionally accepted. But
he premisses from which this conclusion is drawn are too
echnical for discussion in these pages. It must suffice to
state as briefly as possible what this conclusion amounts to,
and to indicate some of the consequences which follow from
ts acceptance.

The fertilized egg gives origin, as we have seen, to the
multitude of cells which build up the body of one of the
higher animals. There are, on the one hand, muscle-cells,
gland-cells, nerve-cells, and other constituents of the various
issues; and there are, on the other hand, the reproductive
cells—ova or sperms, as the case may be. Now, every cell in
he developed animal is a direct descendant of the fertilized
egg. But of all the varied host only the reproductive cells

take any direct share in the continuity of the race. Hereditary transmission is therefore restricted to the germinal substance of these reproductive cells. Trace the ancestry of any cell in the adult body, say a nerve-cell, and you reach the fertilized ovum. Trace back the ancestral line yet further, and you follow a long sequence of reproductive cells, or, at least, of cells which have undergone but little differentiation; but never again will you find, in the course of a genealogy of bewildering length, a nerve-cell. Such a tissue-element is a descendant, but cannot become an ancestor; it dies without direct heirs.

It is universally admitted that the bodily structures are subject to what is termed *modification* under the stress of environing circumstances. The muscles may acquire unusual strength by use and exercise; the nerve-centres may learn certain tricks of behaviour in the course of individual life ; and other structures may be similarly *accommodated* to the conditions which affect them. To such modifications of structure or function in the organs or parts the term *acquired* is primarily applied. The tissues have thus a certain amount of organic plasticity, through which they are adjusted to a range of circumstances varying in extent. They are able to acquire new modes of behaviour. But the cells of which they are composed are off the line of racial descent. They leave no direct heirs. When the body dies the modifications of behaviour acquired by its parts perish with it. Only if in some way they exercise what we may term a homœopathic influence on the germinal substance can the accommodation they have learnt be transmitted in inheritance. By a *homœopathic* influence is here meant one that is of such a nature as to communicate to the germinal substance, the seeds of similar changes of structure or function. And of the occurrence of any such homœopathic influence there is no convincing evidence.

Logically contrasted with the modifications of the tissues, dependent on organic plasticity, are the *variations* which arise from the nature and constitution of the reproductive cells.

How they arise cannot here be discussed. But they are, it is believed, subject to the influence of natural selection, which has guided them, throughout the ages of organic evolution, in the directions they have taken ; disadvantageous variations having been eliminated, and favourable variations surviving in the struggle for existence. Such modes of behaviour as are congenital and are due to hereditary transmission are therefore the outcome of variations which have been selected generation after generation. And the fit adjustment of this congenital behaviour to the needs of life is termed *adaptation*. It is here assumed that modifications of behaviour in one generation are not inherited, and therefore contribute nothing to the store of adaptive behaviour in the next generation.

It must not, however, be supposed that the provisional acceptance of this conclusion involves the denial of all connection of any sort between accommodation and adaptation. When we remember that plastic modification and germinal variation have been working together, in close association, all along the line of organic evolution to reach the common goal of adjustment to the circumstances of life, it is difficult to believe that they have been throughout the whole process altogether independent of each other. Granted that acquired modifications, as such, are not directly inherited, they may none the less afford the conditions under which *coincident variations* escape elimination. By coincident variations I mean those the direction of which coincides with that taken by modification. The survival of an animal depends on its adjustment to the circumstances of its life, no matter how this adjustment is secured. And this survival would in the long run be better secured, we may suppose, where the two methods of adjustment were coincident and not conflicting ; * just as a man who not only acquires by

* Professor Mark Baldwin has applied the term " organic selection " to the result of this interaction (*American Naturalist* for June and July, 1896). Cf. also H. F. Osborn (*Science*, Nov. 27, 1896); August Weismann (Romanes Lecture on " The Effects of External Influences on Development," 1894), and " Germinal Selection," *Monist*, Jan., 1896 ; and the author's " Habit and Instinct," ch. xiv., 1896.

his own exertions a fortune but also inherits one, is better off
than his neighbour, of equal business capacity, who is entirely
dependent on his own exertions. The inheritance of a small
capital may, indeed, make just the difference between success
and failure. Even with it, if he had no power of acquiring
more, he might remain a poor man. Inheritance and acquisition combined may best lead to survival in competition.
Thus modification may supply the conditions under which
coincident variations are favoured, and, given time, to reach
step by step, through natural selection, a fully adaptive level.
If this be so we may accept many of the facts adduced by the
transmissionist in favour of the direct inheritance of acquired
characters, and at the same time interpret them on selectionist
principles.

If, however, acquired characters are not hereditary the
method of natural selection in racial progress is curiously
indirect. Apart from the preservation of their fecundity, the
cells on which the continuity of life, in all the higher animals,
depends, have themselves taken little part in the struggle for
existence. Just as in the forest tree, the firmly implanted
roots, the sturdy stem, and the strong branches have to bear
the stress of the winter storm, that the flowers of spring may
ripen the seeds which contain the potentiality of all this
strength ; so do muscle, sinew, and brain secure the survival
of the animal, that his descendants may carry on the struggle.
One may liken the cellular constituents of the animal to a hive
of bees with fertile drones and queen, and sterile workers. It
is on the exertions of the latter that, in the struggle for
existence, the continued existence of the swarm depends, while
it is by the pairing of the fertile drone and queen that the
continuity of the race is secured. No worker can transmit
the qualities which are so essential to the well-being of the
community. But in the eggs of their sister the queen-mother
these qualities lie dormant. And since the race is one race,
the workers by their exertions contribute indirectly to the
maintenance of those hereditary aptitudes to which they are
unable to contribute directly. For it is essential to bear in

mind that they not only work for their own generation, but they determine the course of heredity. Picture two such communities set in an environment which intensifies the struggle for existence. The one is strong, healthy, and vigorous ; the other in all respects the reverse. The incidence of the battle of life falls mainly on the workers. If they succumb in the one group their fertile queen either perishes, or gives rise to a poor stock, certain in the long run to be eliminated. But the vigorous workers in the other group survive and secure, too, the survival of their queen, who, since she is also their sister, bears, in her ovaries, the good seed from which a new generation of vigorous workers shall be developed. Thus though the sterile bees contribute nothing directly to the heredity of the race, they indirectly determine the direction which that heredity shall take. So, too, in the higher animals, the reproductive cells are the fertile sisters of a host of sterile body cells, on which the main incidence of the struggle for existence falls. Their sterility precludes their directly contributing to the success of future cell-generations ; but in protecting their fertile sisters, the reproductive cells, they are really determining the lines along which the evolution of the race shall continue.

Acquired characters may thus be regarded as the results of those accidents, fortunate or the reverse as the case may be, which happen to the body, and more or less modify its outward form or hidden structure, and its modes of organic behaviour ; but which, as such, have no direct effects for better or worse on the germinal substance. All that the plant or animal can be is due to heredity ; all that it *is*, to heredity and circumstance. Even the ability to yield to circumstance is part of heredity's dower. Fortunate, then, the plant or animal that inherits such definiteness of structure and behaviour as may fit it to its station, together with such plasticity as may enable it to accommodate itself to those changes of environing conditions which may fall to its lot.

One more point must be noticed in connection with this difficult and puzzling subject. The acceptance of the

conclusion that acquired modes of behaviour are not hereditary nowise commits us to the belief that heredity has nothing
whatever to do with them.   Though what is acquired may
not be transmitted, what one may term the acquisitiveness is
unquestionably inherited.   Though this, that, or the other
acquired mode of behaviour may have no direct descendants,
the power of acquiring any one of them under the appropriate
circumstances is handed on as an invaluable legacy.   Just as
the mirror which has reflected a fleeting scene retains no
lasting image of the bygone events, so heredity may retain no
impress of acquired characters ; but just as the mirror keeps
its power of reflecting such scenes, so does heredity transmit
the power of acquiring such characters.   As the leaves of the
oak are renewed each successive spring, so may acquired modes
of behaviour be repeated in each successive generation if only
the requisite conditions recur in due season.

From what has preceded it may, therefore, be inferred that
organic behaviour may arise either through modifications
occurring in the plastic tissues, or through variations having
their origin in the germinal substance.   Broadly speaking,
however, we may regard as predominantly due to adaptation
those congenital modes of behaviour and those organic responses which on their first occurrence are relatively definite
in character, and which are directed to a biological end, for
whose attainment the tissues have had no preparatory training ;
and we may regard as predominantly due to accommodation
those responses which are, so to speak, learnt by the tissues in
the course of individual life.   Both are dependent on heredity,
but in different ways.   What the animal owes to heredity may,
indeed, as I have elsewhere said,* be classified under two
heads.   Under the first will fall those relatively definite modes
of behaviour which fit the animal to deal at once, on their
first occurrence, with certain essential or frequently recurring
conditions of the environment.   Under the second head will
fall the power of dealing with special circumstances as they
arise in the course of a varied life.   The former may be

* " Habit and Instinct," p. 26.

likened to the inheritance of specific drafts for definite needs which are sure to arise in the conduct of life ; the latter to the inheritance of a legacy which may be drawn upon for any purpose as occasion may demand. If the need becomes habitual the animal may, so to speak, instruct his banker to set aside a specific sum to meet it as it arises. But this arrangement is a purely individual matter, dictated by experience, and in no wise enjoined by the original terms of the bequest. And both types are fostered by natural selection which develops (a) such congenital definiteness of response, and (b) such innate plasticity, as are advantageous under the conditions of existence ; uniform conditions tending to emphasize the former, variable conditions the latter.

Difficult as it may be to earmark the items of the organic bequest—to say that, of the sum of energy expended in any given case of organic behaviour, so much is due to a specific draft definitely assigned in heredity for this particular purpose, and so much is contributed from the general legacy of innate plasticity,—it none the less conduces to clear thinking to emphasize the logical distinction between them, so long as it is steadily borne in mind that logical distinction does not imply biological separation. The animal, with all its varied modes of behaviour, is an organic whole, and as an organic whole it has been developed from the fertilized egg. The very same tissues which exhibit congenital modes of behaviour are capable also of acquiring new responses and playing their part in accommodation. We have not one set of organs which are the products of variation and another set which result from modification. Our study would no doubt be simplified if this were the case ; but it is not so. And we must take the animal as we find it, presenting varied behaviour of complex origin. Even the reflex nervous centres, which are concerned in responses so automatic as to suggest a stereotyped structure of distinctively germinal origin, are also, as we saw at the close of the last section, in close touch with those centres of control which are associated with the supreme power of accommodation arising from the possession of consciousness.

CHAPTER II

CONSCIOUSNESS

It is possible that all organic behaviour is accompanied by
consciousness. But there is no direct means of ascertaining
whether it is so or not. This is, and must remain, a matter of
more or less plausible conjecture. We have, indeed, no direct
knowledge of any consciousness save our own. Undue stress
should not, however, be laid on this fundamental isolation of
the individual mind. We confidently infer that our fellow-
men are conscious, because they are in all essential respects
like us, and because they behave just as we do when we act
under its guiding influence. And on similar grounds we
believe not less confidently that many animals are also conscious. But how far we are justified in extending this
inference it is difficult to say. Probably our safest criterion
is afforded by circumstantial evidence that the animal in
question profits by experience. If, as we watch any given
creature during its life-history, we see at first a number of
congenital or acquired modes of behaviour, we may not be
able to say whether they are accompanied by consciousness or
not ; but if we find that some of these are subsequently carried
out more vigorously while others are checked, we seem justified
in the inference that pleasurable consciousness was associated
with the results of the former, and disagreeable consciousness
with those of the latter. When we see that a chick, for
example, pecks at first at any small object, it is difficult to

say, on these grounds, whether it is a sentient animal or only an unconscious automaton ; and if it continued to behave in a similar fashion throughout life, our difficulty would still remain. But when we see that some objects are rejected while others are selected, we infer that consciousness in some way guides its behaviour. The chick has profited by experience. But even this is clearly only a criterion of what we may term *effective* consciousness. There may be sentience which is merely an accompaniment of organic action without any guiding influence on subsequent modes of behaviour. In that case it is not effective ; and whether it is present or not we have no means of ascertaining.

We seem also to be led to the conclusion, both from *a priori* considerations and from the results of observation, that effective consciousness is associated with a nervous system. Its fundamental characteristic is control over the actions, so that some kinds of behaviour may be carried out with increased vigour, and others checked. And it is difficult to see how this can take place unless the centres of control are different from those over which they exercise this influence. If we are to understand anything definite by the guidance of consciousness, we must conceive it as standing apart from and exercising an overruling influence over that which it guides. This is un-questionably an essential characteristic of consciousness, as generally understood by those who take the trouble to con-sider its relation to behaviour ; and though some would seek to persuade us that a mere accompaniment of consciousness can somehow determine the continuance or discontinuance of organic behaviour, it is difficult to see how this can be the case. The accompaniment of air-tremors can no more influence the vibrations of a sounding string than an accompaniment of consciousness can affect the nature of the organic changes in the tentacles of the Sun-dew leaf.

And if, instead of trusting to such general *a priori* con-siderations, we study with attention the conditions under which an animal so behaves as to lead us to infer that it profits by experience, we find that it is not the consciousness that accompanies the behaviour which leads to future guidance, but the consciousness that arises from the results of the behaviour. Let us willingly grant that the newly hatched, and as yet inexperienced chick, when it pecks at a small object is conscious of a visual impression, and is conscious also of movements of its neck and beak. These do not constitute the experience by which it profits. This is provided by the results of the pecking, according as the morsel seized is nice or nasty. We may say, in popular language, that the little bird remembers when it sees a similar object that the former results were pleasant or distasteful, as the case may be ; and that it is through this remembrance that future guidance is rendered possible. But all the evidence that we possess goes to show that the sensory centres, stimulated by what we will assume to be the taste of the morsel, are different from those which are affected by sight, and the movements concerned in pecking. So that the consciousness which is effective in guiding future action is an accompaniment of the stimulation of centres that are different from those concerned in the behaviour over which guidance is exercised. And if this interpretation of the observed facts be correct, it supports the conclusions reached from *a priori* considerations. It seems further to show that, not only is a nervous system necessary for the occurrence of controlled behaviour, but that no little complexity in its intercommunications is essential.

It may be urged that the chick's behaviour which has been selected for purposes of illustration, and the interpretation we have put upon it, throws too much stress on remembrance, so called, and further gives the false impression that all experience must be for *future* guidance. There are surely numberless cases, it will be said, in which nothing of the nature of distinct memory is involved, and in which the guidance of consciousness is exercised at once over present behaviour, without any postponement to the future. Even omitting for the present the former point, the formula implied —that present experience is for future guidance—cannot be accepted in view of the familiar fact that present experience is constantly influencing present behaviour. Practically speaking, this is perfectly true : because, practically, under the term present we include quite an appreciable period of time—say, a few seconds, or even minutes. If we narrow our conception of the present, as is commonly done in philosophical discussions, to the boundary line between past and future, then it will be seen that even the guidance of what in popular speech is called present behaviour is really exercised on the *subsequent* phases of that behaviour. At the risk of some technicality our position must be explained a little more fully. It is assumed that the data of consciousness are afforded by afferent impulses coursing inwards from the organs of special sense, or those concerned in responsive movements. This conclusion rests on such a wide body of psychological inference that it may be accepted without discussion, at any rate for our immediate purpose. The efferent impulses, those which effect the orderly contraction of the muscles, are unconscious ; but when the movement is produced afferent impulses course inwards from the parts concerned in the behaviour, and these may then contribute data to consciousness.

Now let us suppose that a chick, which has been hatched in an incubator, be removed some twelve hours after birth, held in the hands for a few minutes until its eyes have grown accustomed to the light, and placed on a table near some small pieces of hard-boiled egg. Let us watch its behaviour and endeavour to interpret it. We shall have occasion to consider hereafter whether the conscious experience of parents and ancestors is inherited as such ; for the present we will assume that it is not. The chick has to acquire for itself its own experience. A piece of egg catches the eye of the little bird, which then pecks at it, and just fails to seize it. Here is a piece of congenital organic behaviour. Taken by itself one might find it difficult to say whether it is accompanied by consciousness or not, just as one finds it difficult to say whether the closure of the Venus's Fly-trap is conscious. But the subsequent behaviour of the chick leads us to infer that it is a sentient animal ; and we may, therefore, fairly assume that it is sentient from the first. Dividing the course of the observed behaviour into stages, we may say that the first stage is that in which the chick receives a visual stimulus accompanied by a sensation of sight. Upon this there rapidly follows the second stage, when the bird pecks, and its experience is widened by new data of consciousness derived from a group of motor sensations ; and upon this, again, there follows the third stage, when sensations come in from the morsel of egg which the chick touched but just failed to seize. After a pause the chick strikes again. But we have not a mere repetition of the former sequence of stages. The visual stimulus at first fell upon the eye of a wholly inexperienced bird ; now it falls upon the eye of one that has gained experience of pecking and tasting. What we may call the *conscious situation* has completely changed, at all events if we assume that the items of consciousness, including as essential the consciousness of behaviour, do not remain separate and isolated, but have coalesced into a group through association. And in this group the consciousness of behaving is perhaps the most important element in the situation, making it of practical value. What psychologists term the *presentative* visual stimulus, now calls up *re-presentative* elements, motor and gustatory ; and these place the situation in a wholly new aspect. They give it what Dr. Stout terms " meaning." On the second or third attempt the chick seizes and swallows the morsel of egg. Its experience is yet further widened ; and thereafter the situation has other new elements. Later it pecks at some nasty grub ; shakes its head, and wipes its bill on the ground. The conscious situation has for the future become more complex, and the behaviour is henceforth differentiated into that of acceptance and that of rejection, in each case determined by the acquired meaning in the coalescent conscious situation : the sight of a nice piece of egg being one situation, that of a nasty caterpillar another, each associated with its specific behaviour-consciousness. We need not carry the illustration further on these lines : the essential feature is that experience grows by the coalescence of successive increments, and that each increment modifies the situation which takes effect on the *succeeding* phases of behaviour, even if they succeed within the fraction of a second. That is what is meant by saying that present experience is for future guidance. The future need not be remote, but may be so immediate that in popular speech we may say that it is not future but present guidance which is rendered possible.

We may now turn for a moment to the criticism that there are numberless cases in which nothing of the nature of distinct memory is involved. We may now substitute for the word remembrance, which was used above, the more technical term *re-presentation*. Profiting by experience, regarded as a criterion of the presence of effective consciousness, involves re-presentative elements in the conscious situation which carry with them meaning. Let us for the moment assume an ultra-sceptical attitude with regard to any conscious accompaniment. The chick when it pecks, let us say, is an unconscious automaton. It seizes a piece of egg ; this affords an unconscious stimulus, which sets agoing unconscious acts of swallowing ; or it seizes a piece of meal soaked in quinine, which sets agoing unconscious acts of rejection and touches the hidden springs which make the automaton wipe its bill. So far we find no great difficulty. It is when we have to consider subsequent behaviour that a severe strain is felt on this method of interpretation. One can understand an automatic action repeated again and again as often as the stimulus is repeated. But the chick may shake its head and wipe its bill on the mere sight of the quinine-soaked meal, which, on the hypothesis of conscious experience, has already proved distasteful. So that if we accept the unconscious automaton theory we must assume an organic association which closely simulates the conscious association to which our own experience testifies. But the associations which take part in the guidance of behaviour in the chick are so varied and delicate, so closely resemble those which in ourselves imply conscious guidance, that a sceptical attitude throws more strain upon our credulity than the acceptance of the current belief in conscious control. We shall therefore assume that evidence for such coalescent association is also evidence of the presence of effective consciousness.

It may still be said, however, that in selecting an example from so highly organized an animal as a bird, we are taking for granted that a complex case of controlled behaviour may fairly be accepted as a type of more simple cases. Unfortunately the only being with whose power of conscious control we have any first-hand acquaintance is possessed of a nervous system even more complex than that of the chick. Our psychological interpretations are inevitably anthropomorphic. All we can hope to do is to reduce our anthropomorphic conclusions to their simplest expression. The irreducible residuum seems to be that wherever an animal, no matter how lowly its station in the scale of life, profits by experience, and gives evidence of association, it must have some dim remembrance, or, let us now say, some re-presentation, of the results of previous behaviour which enters into and remodels the conscious situation ; that through the re-presentative elements behaviour is somehow guided ; and, further, that the centre of conscious control is different from the centre of response over which the control is exercised.

II.—The Early Stages of Mental Development

We use the phrase " mental development " in its broadest acceptation as inclusive of, and applicable to, all phases of effective consciousness. We shall assume that throughout this development there is a concomitant development of nerve-centres and of their organic connections. And we shall further assume that experience, as such, is not inherited.

The nature of the grounds on which the latter assumption is based must first be briefly indicated. It is commonly asserted that fear of man, the inveterate hunter and sportsman, is inherited by many animals, as is also that of other natural enemies. This is, however, questioned, or even denied, by many careful observers. Mr. W. H. Hudson has an excellent chapter on " Fear in Birds " in his " Naturalist in La Plata," and concludes that fear of particular enemies is, in nearly all cases, the result of experience individually acquired. I have found that pheasants, partridges, plovers, domestic chicks, and other young birds, hatched in an incubator, show no signs of fear in the presence of dog or cat, so long as the animal is not aggressive. It should be mentioned, however, that Miss M. Hunt * asserts that chicks do show inherited fear of the cat. Dr. Thorndike's † observations, on the other hand, support my own, which I have since repeated with the same results. Neither birds nor small mammals show any signs of fear of stealthily moving snakes. My fox terrier smelt, nose to nose, a young lamb which was lying alone in a field. I was close at hand, and could detect no indication of alarm on the part of the lamb till the mother came running up in great excitement. Then the lamb ran off to her dam. Whenever opportunity has arisen, I have introduced young kittens to my fox terrier, and have never seen any sign of inherited fear. He was a great hunter of strange cats, but was trained to behave politely to all birds and beasts within the precincts of my study. It is true that he was on good terms, or at least terms of permissive neutrality, with the kittens' mother. And it may be said that this was inherited ; but such an argument cannot apply in the case of pheasant or lamb.

Here, as throughout our study of animal behaviour in its conscious aspect, we have not only to conduct observations with due care, but to draw inferences with due caution. Douglas Spalding described how newly hatched turkeys showed signs of alarm at the cry of a hawk ; and he inferred that, since this sound was quite new to their individual experience, the alarm was due to the inheritance of ancestral experience of hawks. But since young birds show signs of alarm at any sudden and unaccustomed sound—a sneeze, the noise of a toy horn, a loud violin note, and so forth—the safer inference seems to be that they may be frightened by strange

* *American Journal of Psychology*, vol. ix., No. 1.
† *Psychological Review*, vol. vi , No. 3.

sounds of many kinds. But this does not imply the inheritance of experience, which is essentially a discriminating process. There is no sufficient evidence that a peculiar cry suggests the hawk, of which the progenitors have acquired bitter experience ; nothing to justify the belief that the sound carries with it inherited meaning. And as with hearing, so with sight. Young birds may be frightened by many strange objects. I have seen a group of several species, filled with apparent alarm at a large white jug suddenly placed among them, at balls of paper tossed towards them, at a handkerchief dropped in their midst. It is, in fact, their inexperience which is often the condition of such fear. As Mr. Hudson says : *
" A piece of newspaper carried accidentally by the wind is as great an object of terror to an inexperienced young bird as a buzzard sweeping down with death in its talons."

Until recently it was commonly asserted that birds avoid gaudy but nauseous or harmful insects through the inheritance of experience gained by their ancestors through many generations. But here again the inference seems to have been incautiously drawn. Of the hundreds of young birds I have had under observation, not one has avoided the peculiarly distasteful cinnabar caterpillar, until it had gained for itself experience of its nauseous character. So too of wasps and bees. Only through experience are these avoided. It is true that chicks may shrink from them if they buzz or even walk rapidly towards them. But a large harmless fly will inspire just as much timidity. As the result of careful observations, Mr. Frank Finn † concludes " that each bird has to separately acquire its experience, and well remembers what it has learnt." And with this conclusion my own observations are entirely in accord.

Such is some of the observational evidence on which is based the provisional hypothesis that experience, as such, is not inherited. What, then, is inherited ? Clearly the organic conditions under which experience can be acquired. Since a

---

* " Naturalist in La Plata," p. 88.
† *Journal Asiatic Society of Bengal*, lxvii., part ii., 1897, p. 614.

young bird inherits a tendency to peck at small objects, especially, in the case of some birds such as plovers or partridges, at small moving objects, opportunities are afforded for discrimination in accordance with the results of experience. Since its inherited timidity leads the chick to shrink from many things seen or heard, a wide range of conscious data is supplied. Inheritance provides the raw material of organic behaviour for effective consciousness to deal with in accordance with the results which are its data.

Having thus cleared the ground and laid bare some at least of the assumptions which we accept as foundations on which to build, we may now follow up the line of treatment which was suggested in the first section of this chapter. Remembering that our aim is to understand the influence of consciousness on behaviour—or, in more accurate, if more cumbrous phraseology, the influence of certain nerve-centres which have for their concomitant what we have termed effective consciousness—the questions which present themselves in any given case are : What is the conscious situation which is effective in guidance ? what elements enter into the situation, whence are they derived, and how were they introduced ? how do they take effect in behaviour ?

If it be true that, in many of the lower forms of life, consciousness or sentience, though presumably present in some dim form, is merely an accompaniment of organic behaviour without reaching the level of recognizable effectiveness ; and if, during the development of one of the higher animals from the fertilized ovum, the early stages of organic behaviour are in ·like manner merely sentient ; it follows that, when effective consciousness enters upon the scene (who can say at what exact stage of evolution ?), it finds itself a partner in a going concern. Much organic business is being transacted with orderly regularity ; preparations have been made for more extensive operations ; and energies lying dormant, or expending themselves aimlessly in starts and twitches, await the guidance which shall direct them to higher and wider biological ends. Or, to vary the analogy, consciousness is

the heir to a wide estate, over which he has no control until
he comes of age.   Up to that time the estate is managed in
strict accordance with the dictates of the hereditary bequest.
He may be aware of what is going on, but merely as a spectator without power of interference.   And when he comes into
possession his first business is to gather up the threads.   He
must learn bit by bit how the estate is being managed, that he
may have data for the guidance of his own management
within the wise limits of the hereditary entail.

Now, when a mammal is born, a bird is hatched, an insect
emerges from the chrysalis, we have, if not the beginning, at
any rate a great and sudden extension of the range of effective
consciousness.   In the case of the mammal and bird the
experience gained in the womb or within the egg-shell is
presumably of little value for the wider life upon which an
entrance is made.   It is true that an insect has passed through
a previous stage of active and no doubt consciously guided
existence as a caterpillar.   But we do not know whether the
experience thus acquired is effectual for use in the later imago
stage.   And we may perhaps infer from the extensive re-
modelling of the nervous system, which occurs during the
chrysalis sleep, that this itself serves to break the continuity
of experience.   In any case the newly hatched chick, if it
inherit no experience, and can have gained little of guiding
value in the egg, enters upon a situation which from the
number and variety of the data supplied may well seem to
us bewildering.   If we picture ourselves in such a position,
with sights, sounds, motor sensations, touches, and pressures
raining in upon a virgin experience, we wonder how we should
make a beginning; how we could possibly decide on the first
step towards reducing this multiplicity and diversity to something like unity and order.   And perhaps we wonder how we
ourselves made a beginning when we were pink newly born
babies.

If it may be said without paradox, we never did make a
beginning.   The beginning was made for us.   For we habitually
associate ourselves with the control centres, and regard our

bodies, like our watches, as ours and not us. We wind the bodily watch, and set its hands from time to time ; but we did not make it, and it was already going when heredity handed it to us over the counter of birth. The first step towards reducing the seeming chaos of sensory experience to something like order is not due to the selection by consciousness of this or that element for prominence among the rest, but to the thrusting forward of certain modes of behaviour by the conditions of organic life. The differentiation of the field of vision in the chick is not effected by any conscious determination to fix the attention on that wriggling maggot, but through the congenital response it calls forth. This serves not only to make the grub stand out clearly amid its surroundings, but also to emphasize a motor group, called into vigorous action in the midst of other motor sensations, and, in rapid sequence, to lay stress on a sensation of taste suddenly called into prominence.

Nor, as we have seen, do the organic effects cease here. The functional action of three sensory centres is thus called into play. But they are constituent parts of one nervous system. The direct stimulation of each by nerve impulses from eye, motor organs, and beak, gives temporary predominance to certain sensory data which are termed presentative. But the several centres are connected with each other. And thenceforward, in subsequent stages of experience, the direct stimulation of the visual centre indirectly calls into play the other two, so that the presentation through sight evolves re-presentations of the motor group, and of taste. Hence sentience is not sufficient for guidance ; there must be *consentience* involving the presence of several elements. But these elements must not be regarded as separate save in our analysis ; they form constituent parts of the coalescent situation as a whole, of which alone the chick is presumably conscious, without analysis of detail.

It is just because the chick is a going concern when consciousness comes of age and begins to assume control—just because a wide range of congenital behaviour is part of the organic heritage—that the early stages of the acquisition of experience proceed so rapidly and so smoothly. The animal has not to make and fashion the early conscious situations; it has only to accept them. It has not at first to enforce order on the multiplicity of sensory data raining in upon the conscious centres; it has only to take note of the existing order among them. It has not painfully to learn how to co-ordinate the efferent impulses proceeding to the many muscles concerned in some simple response; it has only to be sensitive to the response as a whole. It has not to select the association of this, that, and the other group of data within a coalescent situation; organic behaviour provides it with pre-determined sequences ready made—sequences which have for generations received the emphatic sanction of natural selection. Congenital tendencies which it has inherited but not acquired determine all its earliest behaviour, determine what elements in the sensory complex shall be thrust into conscious pro-minence, determine in what manner these data shall be asso-ciated; determine, in fact, what salient points in the developing situations shall stand out clearly from the rest, and how these salient points shall be grouped and linked by the connecting threads of association and shall coalesce into effective wholes.

And if in the comparatively helpless human infant the congenital modes of response seem less organized than those of the chick, if there is a larger percentage of random and apparently aimless movements, if the organic management of the bodily estate is less definitely ordered by the terms of the hereditary bequest, if there is more of maternal guidance and fosterage; still the data are provided in a substantially similar way. The situations are indeed destined to become more complex, the distinctions which arise in consciousness are more numerous, the coalescence and association include a wider range and succession of salient points; a longer time is required to become acquainted with the transactions of a business con-ducted in a far greater number of centres: but, at least in the early stages, the data are of the same kind, and are emphasized in the same way. Presentation and re-presentation play a similar *rôle ;* and the chief difference lies in the fact that less stereotyped congenital behaviour is supplemented by some guidance, probably far less than is generally supposed, from those who lovingly minister to the course of infant development.

No attempt can here be made to trace even in outline (an outline which must in any case be imaginary and conjectural) the sequence of situations which marks the course of mental development in its earlier stages. An example may, however, serve to show how the exercise of congenital tendencies may give rise to a new situation, and lead to a further development of behaviour.

I kept some young chicks in my study in an improvised pen floored with newspaper, the edges of which were turned up and supported, to form frail but sufficient retaining walls. One of the little birds, a week old, stood near the corner of the pen, pecking vigorously and persistently at something, which proved to be the number on the page of the turned-up newspaper. He then transferred his attention and his efforts to the corner of the paper just within his reach. Seizing this, he pulled at it, bending the newspaper down, and thus making a breach in the wall of the pen. Through this he stepped forth into the wider world of my study. I restored the paper as before, caught the bird, and replaced him near the scene of his former efforts. He again pecked at the corner of the paper, pulled it down, and escaped. I then put him back as far as possible from the spot. Presently he came round to the same corner, repeated his previous behaviour, and again made his escape.

Now, here the inherited tendency to peck at small objects led, through the drawing down of the paper, to a new situation, of which advantage was taken. The little drama consisted of two scenes, which may be sufficiently described as " the corner of the pen," and " the open way," this being the sequence in experience. Subsequently the first scene was again enacted in presentative terms, and there followed first a re-presentation of scene ii., with its associated behaviour, and then the presentative repetition of this scene. We may

take this as a sample of the nature of a conscious situation
which is effective in guidance. We have seen the nature of
the elements (sensory data, including as essential those supplied by the behaviour itself, with a pleasurable or painful
tone) which enter into such a situation ; we have seen that
they owe their primary origin to direct presentation, but that
they may be subsequently introduced indirectly in re-presentative form ; we have seen that the situation as a whole results
from the coalescence of the data. There only remains the
question how the felt situation takes effect on behaviour.
And to this question, unfortunately, we can give but a meagre
and incomplete reply. All we can say is, that connections
seem to be in some way established between the centres of
conscious control and the centres of congenital response ; and
that through these channels the responsive behaviour may be
either checked or augmented (as a whole or in part), according to the tone, disagreeable or pleasant, that suffuses the
situation. How this is effected we do not fully know.

III.—Later Phases in Mental Development

Some surprise may be felt that in our brief discussion of
the early stages of mental development nothing has been said
of percepts and concepts, nothing of abstraction or generalization. The omission is not only due to a desire to avoid the
subtle technicalities of psychological nomenclature. It is
partly due to the wish not to forejudge a difficult question of
interpretation. Spirited passages of arms from time to time
take place between psychologists in opposing camps, as to
whether animals are or are not capable of forming abstract
and general ideas ; and untrained camp followers hang on the
skirts of the fray, making a good deal of noise with blank-
cartridge. The question at issue turns partly on the definitions of technical terms ; partly, when there is agreement on
this point, on the interpretation to be put on certain modes of
behaviour. Nothing seems at first sight much easier than to
say what we mean by an abstract idea or by a general idea.

We are thinking about colour, which is both abstract and general—abstract, because in itself it is a special quality of visible objects floated off, so to speak, from other qualities, such as hardness and weight, shape and size ; general, because it includes many different colours in one group. Looking up at the bookshelves, we see a volume with a red back. We neglect the shape, the contents, the lettering ; it is the colour with which we are immediately concerned, which forms an important feature in the present thought-situation ; and this is, in virtue of that situation, abstracted from the rest. But a chick a few days old may have acquired experience of several kinds of caterpillars much alike in shape and size ; of which, one kind is ringed with orange and black. And while the others are eagerly seized, caterpillars of this kind are left untouched. It is not the size or the shape which is an effective element in the situation ; it is the peculiar coloration of the cinnabar caterpillars. Now, does the effectiveness of this quality in the stimulus justify the inference that the chick forms an abstract idea of colour ? That clearly depends on our definition of abstract idea, and on our inferences concerning the nature of the chick's mind.

A dog lies dozing upon the mat, and hears a step in the porch without. His behaviour at once shows that this enters into the conscious situation. There is, moreover, a marked difference according as the step has the familiar fall of the master's tread, the well-known shuffle of the irrepressible butcher's lad, or an unfamiliar sound. These several situations are, without question, nicely distinguished. Let us suppose the situation of the moment is introduced by a strange foot-fall. It seems to suggest man ; but this cannot be any particular man, since he is as yet invisible and is a stranger. Does the dog, then, frame a general idea of man ? Does the chamois do so when, bounding across the snow field, he stops suddenly on scenting the distant footprints of a mountaineer ? Do you do so when you hear the bleating of an invisible lamb in the meadow behind yonder wall ? Here, again, the answers we give to these questions depend partly on the exact meaning

of the term " general idea ; " partly on our interpretation of
what passes through the mind of the being concerned. We
have sought, so far, rather to avoid than to answer these
questions. We seem to be on safe ground so long as we
content ourselves with saying that the orange and black of
the cinnabar caterpillar, the strange footfall, or the trail of
the mountaineer, enter as effective elements into the immediate
conscious situation.

But when we pass to the higher phases of mental development we can no longer wholly ignore such questions. When
we are dealing with intellectual human beings, there can
be no doubt that they at least are capable of framing, with
definite intention, and of set purpose, both general and abstract
conceptions. And how do they reach these conceptions ? By
reviewing a number of past situations, analyzing them, intentionally disentangling and isolating for the purposes of
their thought certain elements which they contain, and
classifying these abstracts under genera and species—that is
to say, into broader and narrower groups. The primary and
proximate object of this process is to reach a scheme of thought
by which the scheme of nature, as given in experience, can be
explained. And, no doubt, underlying this primary object is
the purpose of guiding future behaviour in accordance with
the rational scheme which is thus attained. Man is sometimes
described as *par excellence* the being who looks before and
after. All his greatest achievements are due to his powers of
reflection and foresight.

What share the symbolism of speech takes in the process
briefly indicated in the last paragraph is the subject of much
discussion. Without going so far as to urge that the very
beginnings of reflective thought are inexplicable without its aid,
it may be accepted as obviously true that words are a great
assistance. They may be regarded as intellectual pegs upon
which we hang the results of abstraction and generalization.
It may be said that we often think in pictures or images, and
not in words ; but the more abstract and general our thought,
the more it is dependent on the symbolic elements.

We may say, then, that the higher phases of mental development are characterized by the fact that the situations contain the products of reflective thought, presumably absent in the earlier stages ; they are further characterized by a new purpose or end of consciousness, namely, to explain the situations hitherto merely accepted as they are given in pre-sentation or re-presentation; they require deliberate attention to the relationships which hold good among the several elements of successive situations ; and they involve, so far as behaviour is concerned, the intentional application of an ideal scheme with the object of rational guidance. We shall follow Dr. Stout in terming this later stage of mental development the *ideational stage* ; and in speaking of the simpler situations considered in the preceding section as belonging to the *perceptual stage*.

It should be observed that we are not attempting to determine just where, in the scale of organic existence, the line between the perceptual and the ideational stages of mental development is to be drawn. We are certainly very far from asserting that the one does not give rise to the other in the course of an evolution which is orderly and progressive. We are merely contrasting the rational guidance of effective consciousness at its best with the earlier embryonic condition out of which it has arisen by natural genesis. In doing this we have been forced to make some reference to the difficulties of technical nomenclature. And some further reference is necessary lest our point of view be misunderstood.

We shall regard these abstract and general ideas as the products of an intentional purpose directed to the special end of isolating the one and of classifying the other ; we shall reserve the term *rational* for the conduct which is guided in accordance with an ideal scheme or deliberate plan of action ; while for behaviour to the guidance of which no such reflection and deliberation seems to have contributed we shall reserve the term *intelligent*. If, for example, the rejection of a cinnabar caterpillar by the chick is the direct result of experience through the re-presentation in the new situation

of certain elements introduced during the development of
a like situation, we shall call it an intelligent act. But if we
have grounds for supposing that the situation is reflectively
considered by the chick in relation to an ideal and more or
less definitely conceived plan of action which is (perhaps
dimly) taking form in its mind, we shall regard it as so far
rational. And so, too, in other cases of animal behaviour. Now,
with regard to the control through which consciousness is
effective in the guidance of behaviour, it is necessary, in view
of these considerations, to distinguish its intelligent from its
rational exercise. And this is of importance since we generally
speak of control in the latter sense in reference to human
conduct. Intelligent control (on the perceptual plane) is due
to the direct operation of the results of experience without
the intervention of any generalized conception or ideal. Iu
rational control (on the ideational plane), such conceptions
and ideals exert a controlling influence. If, to prevent a boy
sucking his thumb we administer bitter aloes, we trust to
intelligent control through the immediate effects of experience ;
but if he be induced to give up the habit because it is babyish,
he so far exercises rational control. What we call self-control
is of this type. Only one more distinction need be drawn.
Intelligent behaviour, founded on direct association gained
through previous experience, we shall attribute to *impulse* ; but
for rational conduct, the outcome of reflection and deliberation,
we seek to ascertain the *motive*. In human affairs our motives
are referred to certain categories each of which presupposes an
ideal scheme, prudential, æsthetic, ethical, or other. To act
from motive and not from impulse is to act deliberately,
because we judge the action to be expedient, seemly, or right,
as the case may be. If, then, we contrast the lower perceptual
stages of mental evolution with the higher ideational phases,
the former includes behaviour due to impulse ; but from it
conduct due to motive is excluded.

IV.—THE EVOLUTION OF CONSCIOUSNESS

The origin of consciousness, like that of matter or energy, appears to be beyond the pale of scientific discussion. The appearance of effective consciousness on the scene of life does indeed seem to justify the belief in the prior existence of sentience as the mere accompaniment of organic behaviour. *Ex nihilo nihil fit.* And since effective consciousness must, on this principle, be developed from something, it is reasonable to assume that this something is pre-existing sentience. Again, we may assume that this sentience is a concomitant of *all* life-processes, or only of some. But we have no criterion by which we can hope to determine which of these alternatives is the more probable.

We appear, however, at all events to have evidence that when effective consciousness does enter on the scene and play its part in the guidance of behaviour, its progress is, in technical phraseology, marked by that differentiation of conscious elements, and that integration of these differentiated items, which are seemingly the correlatives of the differentiation and integration of nervous systems. There is thus, presumably, a progressive development of orderly complexity in the conscious situations of which controlled or guided behaviour is the outcome. And when this has reached a certain stage—what stage it is most difficult to determine—the relationships, at first implicit in the conscious situations, as they naturally arise in the course of experience, begin to be rendered explicit with the dawn of reflection. Intentional abstraction and generalization to which data are afforded by the reiterated emphasis in experience of the salient features in successive situations, supply new elements to the more highly developed situations of rational life. Ideal schemes and plans of action, the products of reflection and foresight, take form in the mind and enter into the conscious situation. And the intelligent animal, hitherto the creature of impulse, guided only by the pleasurable or painful tone which gives colour to experience, becomes a rational being, capable of judging how far his own behaviour and that of others is conformable to an ideal.

If, then, we were asked to characterize in the briefest possible terms the stages of conscious evolution, we should say that in the first stage we have consciousness as accompaniment ; in the second, consciousness as guide ; in the third, consciousness as judge. And if we were pressed to apply distinctive terms to these three, we should adopt St. George Mivart's term *consentience* for the mid-phase, and speak of mere sentience in the first stage ; consentience in the second ; and consciousness, with restricted signification, in the third and highest stage. Such a distinction in terms is, however, a counsel of perfection, and we shall not attempt to preserve it in the following pages, in which the word " consciousness " will be used in a comprehensive sense.

CHAPTER III

INSTINCTIVE BEHAVIOUR

I.—Definition of Instinctive Behaviour

There are probably few subjects which have afforded more material for wonder and pious admiration than the instinctive endowments of animals. "I look upon instinct," wrote Addison in one of his graceful essays, "as upon the principle of gravitation in bodies, which is not to be explained by any known qualities inherent in the bodies themselves, nor from any laws of mechanism, but as an immediate impression from the first Mover and the Divine Energy acting in the creatures."[*] In like manner Spence said : "We may call the instincts of animals those faculties implanted in them by the Creator, by which, independent of instruction, observation or experience, and without a knowledge of the end in view, they are all alike impelled to the performance of certain actions tending to the well-being of the individual and the preservation of the species."[†] According to such views, instinct is an ultimate principle the natural genesis of which is beyond the pale of explanation. But similar views were, at the time these passages were written, held to apply, not only to animal behaviour, but also to animal structure. The development of the stag's antler, or of the insect's wing, was also regarded as "an immediate impression from the first Mover and the Divine Energy acting in the creatures." This view, however, is, neither in the case of

[*] *Spectator*, No. 120.

[†] Kirby and Spence, "Introduction to Entomology,' Letter xxvii. p. 537 (7th Edit., 1858).

structure nor in the case of behaviour, that enteitained by modern science. It is indeed an expression of opinion concerning the metaphysics of instinct. Leaving the question of ultimate origin precisely where it stood in the times of Addison and of Spence, modern science seeks to trace the natural antecedents of all natural phenomena, and regards structure and behaviour alike as the products of evolution, endeavouring to explain the manner of their genetic origin in terms of progressive heredity.

Omitting, therefore, all reference to problems which, however important, are beyond the limits of scientific inquiry,* we may take as a basis for further discussion Spence's definition, according to which the instincts of animals are those faculties by which, independent of instruction, observation, or experience, and without a knowledge of the end in view, they are all alike impelled to the performance of certain actions tending to their own well-being and the preservation of the species.

Let us first consider the reference of instinctive actions to a *faculty* by which animals are said to be impelled to their performance. Paley also defined instinct as "a *propensity* prior to experience." And unquestionably in the popular conception it is usual to attribute instinctive acts to some such conscious cause. But it will be more convenient, for the present, to consider instinctive behaviour from the objective point of view, as it is presented to our observation ; we may then proceed to the further consideration of the conscious concomitants which may be inferred. From the objective point of view, therefore, we may agree with Professor Groos, who says † that "the idea of consciousness must be rigidly excluded from any definition of instinct which is to be of practical utility," since " it is always hazardous in scientific investigation to allow an hypothesis which cannot be tested empirically." In this we have the support of Dr. and Mrs. Peckham, whose studies of the life-histories of spiders and wasps are models of

---

* Cf. supra, p. 18.

† "The Play of Animals," translated by Elizabeth L. Baldwin, p. 62.

careful and patient investigation. "Under the term Instinct," they say, "we place all complex acts which are performed previous to experience, and in a similar manner by all members of the same sex and race, leaving out as non-essential, at this time, the question of whether they are or are not accompanied by consciousness." *

It may be said, however, that some reference to the conscious aspect of instinctive behaviour is implied by saying that the acts are performed without instruction or experience. But the reference at present is wholly negative. We may say, as the result of observation, that instinctive acts are performed under such circumstances as exclude the possibility of guidance in the light of individual experience, and render it in the highest degree improbable that there exists any idea of the end to be attained. But this is a very different position from that of asserting the presence of a positive faculty or propensity which impels an animal to the performance of certain actions. This it is which, from the observational point of view, is unnecessary. For the reference of a given type of observed behaviour to a " propensity " so to behave or to a " faculty " of thus behaving, is no more helpful than the reference of the development of any given type of structure to a " potentiality " so to develop. We may, therefore, without loss of precision, simplify Spence's definition by stating that instinctive behaviour is independent of instruction and experience, and tends to the well-being of the individual and the preservation of the species.

Let us next consider the clause which affirms that instinctive behaviour is prior to experience. This is well in line with the distinction now drawn by biologists between congenital and acquired characters. It refers them to the former category, and implies that the organic mechanism by which they are rendered possible is of germinal origin. This is not, however, universally admitted. Professor Wundt, for example, approaching the subject from the point of view afforded by the study of man and the higher animals, gives to the term a wider meaning,

* George W. and Elizabeth G. Peckham, "On the Instincts and Habits of the Solitary Wasps," p. 231.

and so defines instinct as to include acquired habits. "Move-ments," he says,[*] "which originally followed upon simple or compound voluntary acts, but which have become wholly or partly mechanized in the course of individual life, or of generic evolution, we term *instinctive* actions." In accordance with this definition, instincts fall into two groups. Those "which, so far as we can tell, have been developed during the life of the individual, and in the absence of definite individual influences might have remained wholly undeveloped, may be called *acquired* instincts." They have become instinctive through repetition. "To be distinguished from these acquired human instincts are others which are *connate*." Now, there can be no question that behaviour which has become habitual through frequent repetition is frequently, in popular speech, described as in-stinctive. We hear it said that the experienced cyclist guides his machine instinctively. And the word is similarly used in many like cases. But we shall find it conducive to precision and clearness of thought to emphasize the distinction between what is acquired in the course of life and what is congenital in the race. And to this end we shall regard behaviour which has "become mechanized in the course of individual life" as due to acquired habit, reserving the term *instinctive* for such behaviour as is independent of individual experience. We shall, in short, so far accept Spence's definition.

In this definition, as in those of the majority of naturalists, it seems to be further implied that instinctive behaviour is of a relatively definite kind, though it is no doubt subject to such variation as is found in animal structure and organization. Mr. Rutgers Marshall, however, in a recent work,[†] protests against any such implication, and urges that "this variableness is so wide that definiteness of reaction cannot for a moment be used as a differentia in relation to instinct without narrowing our conception of the bounds of instinct in a manner to be deplored." "The actions," he says, "connected with the preparation for self-defence, those connected with protection of

---

[*] "Lectures on Human and Animal Psychology," pp. 388, 397, 399.
[†] "Instinct and Reason," pp. 90, 92.

the young, with nest-building, with migration, etc., these actions are surely to be classed as instinctive ; and yet they are exceedingly variable and unpredictable in detail ; all that we can predict is the general trend of the varying actions which result from varying stimuli under varying conditions, and which function to some determinate biological end."

Mr. Marshall then proceeds to argue that we are " warranted in speaking of the ethical instincts, of the patriotic instincts, of the benevolent instincts, and of the artistic instincts ; " and thus leads up to the position, to be further elaborated in his work, that there exists in man a religious instinct which has fulfilled a function of biological value in the development of our race. Now, here again there is much in popular usage of the words *instinct* and *instinctive* which lends support, for what it is worth, to Mr. Marshall's very broad conception of the range of instinct. Again and again we hear, in the pulpit and elsewhere, of the religious instinct ; we hear, too, of the benevolent, patriotic, and artistic instincts, and more besides. But what we are endeavouring to define is a type of behaviour which, as such, is prior to instruction and experience. Can we affirm that patriotic and religious behaviour conforms to such a type ? Is it unquestionably congenital and not acquired ? If we are forced to give negative answers to these questions we must regard Mr. Marshall's conception of instinct (one inclusive of multifarious tendencies which have a biological value) as too broad and too vague to be of any service to us at this stage of our study of animal behaviour.

What, then, shall we understand by Spence's phrase that instinct involves the performance of " certain actions " ? And how far shall we accept it ? We shall take it as implying so much definiteness of behaviour as renders instinctive acts susceptible of scientific investigation, and in this sense shall accept it with some modification of phraseology. We shall freely admit, however, the existence of variations of instinctive behaviour analogous to variations in animal structure. It is the occurrence of such variations that renders the natural selection of instinctive modes of behaviour conceivable. We shall also admit some, nay much, variation in detail. Take, for example, two of the cases which Mr. Marshall cites—nest-building and migration. Both involve, not merely a simple response to a given stimulus, but a complex sequence of actions. In detail there may be much variation even among members of the same species. And yet, can it be questioned that the behaviour as a whole is in each case relatively definite ? May we not even say that it is remarkably definite ? May we not even go further, and assert that only on the assumption that instinctive behaviour is relatively definite, can we regard it as a subject for scientific investigation, and can we hope to distinguish it from other modes of behaviour ?

The next point for consideration in Spence's definition, which we have taken as our text, is his characterization of instinctive acts as " tending to the well-being of the individual and the preservation of the species." Here we have Mr. Marshall with us, for he too lays stress on the fact that instinctive behaviour has reference to a definite biological end. But in saying that the biological end is *the* objective mark of an instinct,* he seems to be in error. Because, in the first place, there are other " objective marks," and because, in the second place, this objective mark is not restricted to instinctive behaviour. According to Spence, a further characteristic of instinctive acts is that they are independent of instruction or experience ; and this serves to differentiate them from other modes of behaviour which are also subservient to a biological end. Intelligent behaviour, not less than that which we term instinctive, has reference to a biological end. Many intelligent acts, for example, have for their object the well-being of the individual ; many subserve race preservation ; these bear, every whit as much as instinctive acts, the " objective mark " which Mr. Marshall regards as characteristic of instinct. And if we turn to his subjective criterion—the absence of any conception of the biological end which the behaviour subserves—Mr. Marshall's position is equally untenable. There are thousands of acquired modes of behaviour, dependent on instruction or

* " Instinct and Reason," p. 91.

experience, in which there is, on the subjective side, so far as we can judge, no conception of the biological end to be attained. What can the animal in the early stages of intelligence know of biological ends ? Mr. Marshall's subjective criterion applies just as much to a wide range of intelligent behaviour as it does to instinctive actions.

In accepting, therefore, Spence's statement that when animals behave instinctively they perform, without a knowledge of the end in view, certain actions tending to their own well-being and the preservation of the species, we must take it in connection with the preceding limitation, remembering that they are also performed without instruction and experience.

A further point for very brief consideration is suggested by the phrase in which Spence says that animals are *all alike* impelled to the performance of certain actions. As it stands it is too sweeping and general. Still, we do require some explicit statement of the facts which he had in mind when he wrote the words "all alike." And we find it with sufficient exactness in Dr. Peckham's definition, where he comprises under the category of instinctive behaviour " all complex acts which are performed previous to experience, and *in a similar manner by all members of the same sex and race*." This places congenital behaviour in line with morphological structure as a subject for comparative treatment.

One more question remains. What shall we understand by "complex acts " ? In the first place, it is well to restrict the term instinctive to *co-ordinated* actions ; and this implies the presence of nerve-centres by which the co-ordination is effected. We thus exclude the organic behaviour of plants, since there is no evidence in the vegetable kingdom of co-ordinating centres. In the second place, the co-ordination is, as we have seen, congenital, and not acquired in the course of individual experience. Young water-birds, and indeed young chicks, as soon as they are born, and have recovered from the shock of birth, can swim with definite co-ordination of leg movements. Here the definiteness is not only congenital, but *connate*, if we use the latter term for an instinctive activity which is performed at or very shortly after birth. On the other hand, young swallows cannot fly at birth ; they are then too immature, and their wings are not sufficiently developed. But when they are some three weeks old, and the wings have attained functional size and power, little swallows can fly with considerable if not perfect skill. The co-ordination is congenital, for it is not acquired in the course of individual experience ; but it is not connate, since it is not exhibited at or shortly after birth. The term *deferred* may be applied to such congenital activities as are thus carried out when the animal has undergone a certain amount of further development after birth.

In the third place, it is customary to distinguish between such reflex actions as have already been briefly exemplified,* and instinctive behaviour. It is, however, by no means easy, if indeed it be possible, to draw any sharp and decisive line of demarcation. Instinct has indeed been well described by Mr. Herbert Spencer as compound reflex action ; hence the distinction between instinctive and reflex behaviour turns in large degree on their relative complexity. It would seem, however, that whereas a reflex act—such as the withdrawal of the foot of a sleeping child when the sole is tickled—is a restricted and localized response, involving a particular organ or a definite group of muscles, and is initiated by a more or less specialized external stimulus ; instinctive behaviour is a response of the animal as a whole, and involves the co-operation of several organs and of many groups of muscles. Partly initiated by an external stimulus or group of stimuli, it is also, seemingly, determined in part, in a greater degree than reflex action, by internal factors which cause uneasiness or distress, more or less marked, if they do not find their normal instinctive satisfaction. This point, however, may be more profitably discussed in connection with the conscious aspect of instinct. If, then, we say that reflex acts are local responses of the congenital type due to specialized stimuli, while instinctive activities are matters of more general behaviour, usually

* Chapter I., Section V.

involving a larger measure of central (as opposed to local or ganglionic) co-ordination, and due to the more widely-spread effects of stimuli in which both external and internal factors co-operate, we shall probably get as near as is possible to the distinction of which we are in search. But it must be remembered that there are cases in which the distinction can hardly be maintained.

We are now in a position to define instinctive behaviour as comprising those complex groups of co-ordinated acts which are, on their first occurrence, independent of experience ; which tend to the well-being of the individual and the preservation of the race ; which are due to the co-operation of external and internal stimuli ; which are similarly performed by all the members of the same more or less restricted group of animals ; but which are subject to variation, and to subsequent modification under the guidance of experience.

II.—INSTINCTIVE BEHAVIOUR IN INSECTS

Since instinctive behaviour is, by definition, independent of experience, and since the animals which act instinctively are also, in many cases, able to act intelligently, it is clear that, apart from hereditary variations, we must expect to find acquired modifications of instinct. As Huber said of bees, their instinctive procedure often indicates "a little dose of judgment." It is, indeed, exceedingly difficult, as a matter of observation, to distinguish between hereditary variation and acquired modification. For the *rôle* played by these two factors in any given behaviour can only be determined if the whole life-history of the individual be known, and if there be opportunities for comparing it with the complete life-histories of other members of its race. And this is seldom possible.

These considerations must be borne in mind as we proceed to a brief study of some of the instinctive modes of behaviour in insects.

Dr. and Mrs. Peckham's investigations on the instincts and

habits of the solitary wasps have been described in a volume *
worthy to be placed by the side of Fabre's "Souvenirs." Their
descriptions seem to glow with the warm sunshine, and are
redolent of the fresh air which afforded the conditions under
which the observations were conducted. We can but regret
that, in extracting from their bright pages some of the salient
facts, the natural delicacy and grace of their treatment must
be lost. For we can only give the dry skeleton which they
have clothed with the flesh of lively detail. They enumerate
the following primary modes of instinctive behaviour :—

1. Stinging.
2. Taking a particular kind of food.
3. Method of attacking and capturing prey.
4. Method of carrying prey.
5. Preparing nest, and then capturing prey, or the
reverse.
6. The mode of taking prey into the nest.
7. The general style and locality of the nest.
8. The spinning or not spinning of a cocoon, and its
specific form when one is made.

When the young *Pelopæus* emerges from the pupa-case
and gnaws its way out of the mud cell, with limp and flaccid
wings, it responds to a touch by well-directed movements of
the abdomen with thrusts of the sting, as perfect as those of
the adult. There is clearly no opportunity here for either
instruction or experience to afford any intelligent guidance.
Stinging is an instinctive act. And it is an act of which great
use is made in the capture of prey which shall serve for food
to the young—it has a biological end. But the wasps of
different species do not have to learn by experience what prey
to attack. It is by instinct, too, that they take their proper
food-supply, one caterpillars, another spiders, a third flies or
beetles. So deeply seated, indeed, is the hereditary preference,
that no fly-robber ever takes spiders, nor will the capturer of
spiders change to caterpillars or beetles. Some keep to a few

* "On the Habits and Instincts of the Solitary Wasps," by George W.
and Elizabeth G. Peckham (1898).

species or genera, while *Philanthus punctatus* preys chiefly or entirely on bees of the genus *Halictus*.

Romanes * thought that the manner of stinging and paralyzing their prey might " be justly deemed the most remarkable instinct in the world." Spiders, insects, and caterpillars are stung, he says, "in their chief nerve-centres, in consequence of which the victims are not killed outright, but rendered motionless ; they are then conveyed to a burrow and, continuing to live in their paralyzed condition for several weeks, are then available as food for the larvæ when these are hatched. Of course the extraordinary fact which stands to be explained is that of the precise anatomical, not to say also physiological knowledge which appears to be displayed by the insect in stinging only the nerve-centres of its prey." Eimer † thought that it " is absolutely impossible that the animal has arrived at its habit otherwise than by reflection upon the facts of experience." "At the beginning," he says, "she probably killed larvæ by stinging them anywhere, and then placed them in the cell. The bad results of this showed themselves ; the larvæ putrified before they could serve as food for the larval wasps. In the mean time the mother wasp discovered that those larvæ which she had stung in particular parts of the body were motionless but still alive, and then she concluded that larvæ stung in this particular way could be kept for a longer time unchanged as living motionless food."

Now, since these wasps, when they have stored their nests and laid an egg on one of the victims, close it up once and for all, and take no further interest in it or its contents, there seems no opportunity, at any rate in the existing state of matters, for the acquisition of that experience on which Eimer relied. But both his explanation and Romanes's difficulty are based on the following assumptions : first, that the victims are instinctively or habitually stung in the chief nerve-centres ; secondly, that when thus stung they are not killed but remain paralyzed for weeks ; and thirdly, that the marvellously definite

* "Mental Evolution in Animals," p. 299.

† "Organic Evolution," translated by J. T. Cunningham, p. 280.

and delicate instinctive behaviour is in direct relation to the
uniform result of prolonged paralysis and consequent preservation of the food in the fresh state.   But Dr. Peckham's careful
observations and experiments show that, with the American
wasps, the victims stored in the nests are quite as often dead
as alive ; that those which are only paralyzed live for a varying number of days, some more, some less ; that wasp larvæ
thrive just as well on dead victims, sometimes dried-up, sometimes undergoing decomposition, as on living and paralyzed
prey ; that the nerve-centres are not stung with the supposed
uniformity ; and that in some cases paralysis, in others death,
follows when the victims are stung in parts far removed from
any nerve-centre.   "We believe," he says, "that the primary
purpose of the stinging is to overcome resistance, and to prevent the escape of the victims, and that incidentally some of
them are killed and others are paralyzed."

If, therefore, as will probably be shown to be the case,
these conclusions are found to be generally true for this interesting group of insects, the mystery of "the precise anatomical, not to say also physiological knowledge which appears
to be displayed" by these wasps turns out to be one of our
own fabrication.   It melts away in the light of fuller and
more searching investigation.

It must not be supposed, however, from what has been
said, that the behaviour in the act of stinging is altogether
indefinite.   On the contrary, each species proceeds in a relatively definite manner with some variation or modification of
method.   *Philanthus punctatus*, for example, stings the bees,
on which she preys, under the neck, and the thrust is at once
fatal.   Dr. Peckham further notes that he was only successful in getting the wasps to sting when they were hunting ;
those that had not yet begun to store the nests paid no
attention to the bees.   This is an example of that internal
factor to which reference was made in the last section.
Marchal observed that *Cerceris ornata* runs the end of her
abdomen along the under surface of the thorax of the bee, and
delivers her thrust at the division of the segments—that is,

where the sting can enter. The action does not imply any physiological knowledge. In general she begins at the neck. Spiders are usually, but not always, stung on the ventral surface. To give but one more example, Dr. Peckham observed in three cases the procedure of *Ammophila urnaria* which preys on caterpillars, and often, after stinging, bites the neck in several places, this process being termed malaxation. In three observed captures, all the caterpillars being of the same species and alike in size, the thrusts were given on the ventral surface near the middle line, between the segments. In the first, seven stings were given at the extremities (there being thirteen segments), the middle segments being left untouched, and no malaxation was practised. In the second,

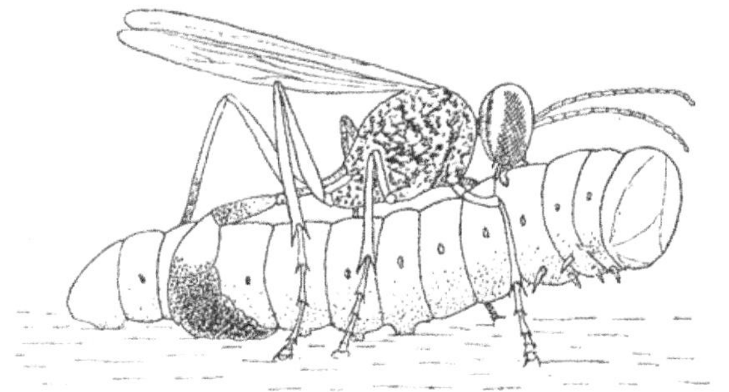

Fig. 11.—Solitary Wasp stinging Caterpillar (after Peckham).

seven stings were again given, but in the anterior and middle segments, followed by slight malaxation. In both these cases the first three thrusts were in definite order, behind the third, the second, and the first segments successively. In the case of the third caterpillar, only one thrust was given, between the third and fourth segments—that is to say, in the position of the first stab in the other cases,—and after this one thrust there was prolonged malaxation. Of fifteen stored caterpillars examined, some lived only three days, others a little longer, while a few showed signs of life at the end of a fortnight. In more than one instance the second of the two caterpillars stored in each nest died and became discoloured before the first one was entirely eaten. The larva under such circumstances ate it with good appetite, and then spun its cocoon as if nothing unpleasant had occurred.

The mode of carrying their booty is in these wasps instinctive, and relatively uniform. *Ammophila urnaria* grasps the caterpillar, near the anterior end, in her mandibles, and carries or drags it beneath her legs, walking forwards. It is

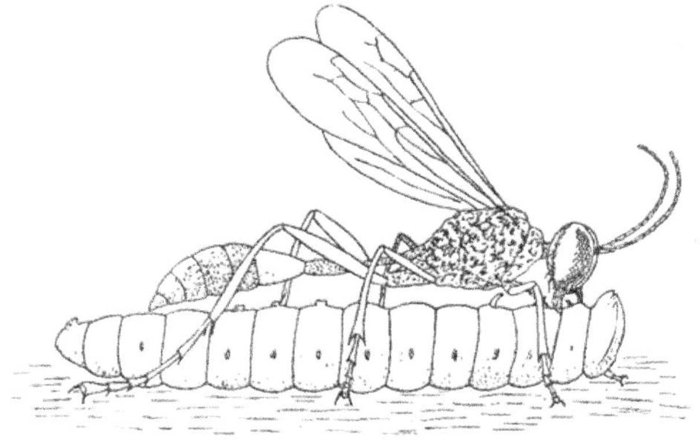

Fig. 12.—Solitary Wasp dragging a Caterpillar to its Nest (after Peckham).

generally but not always with the ventral surface uppermost. *Pompilus* takes hold of her spider anywhere, but always drags it over the ground, walking backwards. *Oxybelus* clasps her fly with her hind legs; *Bembex* with the second pair. Each works after her own fashion in a way that is relatively uniform for each species.

The general style of the nest, its mode of construction, and its method of closure, are always performed, says Dr. Peckham, by each species in a similar manner, not indeed in circumstantial detail, but quite in the same way in a broad sense. Variation or modification is always present, but the tendency to depart from a nest of a given type is not excessive. Some dig in the ground curved tunnels, with or

without one or more chambers. Others bore into decaying
wood ; others use straws, or make tunnels in bramble stems ;
while the mud-daubers build cells in which to store the
food and lay the egg. This is sometimes deposited on the
first, sometimes on the last, sometimes on some intermediate
victim, but generally in much the same place and position.
Ammophila, for instance, lays it on the side of the sixth or
seventh segment—that is to say, in about the mid position.

Some species first capture their prey, and then make the
nest in which it is to be entombed. Others first prepare the
nest, and then carry or drag their prey to it—often from considerable distances—quite irrespective of what seems to us
the more appropriate method of the two under the particular
circumstances of the case. And the way in which the victim
is dragged into the nest is similarly a matter of inheritance.
Each way is characteristic of the species concerned, and would
be an important part of any definition of the animal based
upon its modes of behaviour. For example, a *Sphex* places
her grasshopper just at the entrance of the nest, which she then
enters herself before dragging in her prey by the antennæ.
When the wasp was in the hole, Fabre moved the victim a
little way off ; the wasp came out, brought the grasshopper to
the entrance as before, and went in a second time. This was
repeated about forty times, each time with the same result,
until the patience of the naturalist was exhausted, and the
persistent wasp took her booty in after her appropriate fashion.
She must place the grasshopper close to the opening ; she
must then descend and examine the nest, and, after that, must
drag it down. Nothing less than the performance of these
acts in a certain order satisfies her instinctive impulse.

In a private letter, from which he kindly allows me to
quote, Dr. Peckham says : "We have recently made some
experiments on this wasp (*Sphex ichneumonea*). First we
allow her to carry in her prey undisturbed, to see how far she
was faithful to the traditions of her ancestors, and to observe
her normal methods. On the next day, when she had placed
her grasshopper just at the opening of the nest, and while she

was below, we drew it back to a little distance. She came out, and we both repeated our operations four times—she running down into the nest, always after getting the grasshopper into position, and we as regularly drawing it away. The fifth time she changed her plan, seized it by the head and backed into the nest with it. The next day, at the fourth trial, she straddled it and walked head first into the nest with it ; and on the fourth day, at the eighth trial, she backed in with it as on the second day." These interesting observations show that the wasp has sufficient intelligence to modify her procedure in accordance with an unwonted situation. The "consecutive necessity," as it has been termed, has a potent influence, but is not absolute.

Fabre notes a case of similar consecutive necessity in the case of the mason bee, *Chalicodoma*. If while a bee is provisioning its nest with honey and pollen the structure be destroyed, she sometimes breaks open a completed cell, and, having done so, goes on bringing more provision, though the cell already contains a sufficient store of food ; and only when she has completed the superfluous storing does she deposit her egg and seal up the cell. So, too, when the cell is removed in an early stage of construction, and another completed cell already partially stored is substituted, the bee, instead of simply adopting the new cell, goes on building until the cell is as much as one-third beyond the usual height ; then, and not till then, does she proceed in due course to the next stage of the instinctive procedure, the provisioning of the cell.

From our general knowledge of animal nature, we should expect to find parasitic forms ready to take advantage of the material stored by such insects as the solitary wasps and the mason bees. It is said that *Chalicodoma* provides nourishment to the larvæ of some sixteen unbidden guests. A parasitic bee (*Stelis nasuta*) breaks open a closed cell, and, after depositing its eggs, seals it up again with mortar. Since her eggs and larvæ develop more rapidly than those of the mason bee, they are first served with the store of provision, while the rightful owner is done out of its inheritance. By a curious act, of what appears to us like retributive justice, these parasitic larvæ sometimes fall a prey to another parasite, also a hymenopterous insect named *Monodontomerus*, the larvæ of which prey on the young of both bees.   Another genus of the same family, *Leucopsis* (Fig. 13, F), also succeeds in piercing with its ovipositor, at a suitable spot, the walls of the *Chalicodoma* cell, and suspends its curious hooked egg (Fig. 13, G) on the delicate cocoon within which the chrysalis lies.   Fabre found in some cases as many as five of these parasitic eggs on a single cocoon.   But he never found more than one larva in any cell that he examined.   The following is an epitome of his conclusions and inferences.   From the parasitic egg is hatched a minute arched grub, with relatively large head and mandibles, and provided with a number of bristles, which aid it in progression (Fig. 13, H).   It does not, however, at once attack the bee larva, but makes a series of excursions, the object of which is to reach and destroy any other parasitic eggs.   This was not actually observed, but the eggs were found to have been destroyed, and there was seemingly no other means of destruction under the conditions maintained.   The larva, this done, changes its skin and takes on a new form, destitute of bristles, with a very small head and minute mandibles (Fig. 13, I). In this new form it attacks the *Chalicodoma* larva, making a very small incision, through which the juices of the host are transferred to the guest without further injury to the grub.   It is interesting to note that, if the facts are accurately described and the inferences are correct, there are associated with two types of instinctive behaviour two distinct types of structure.   The creature can have no conscious control over its structural development, and there is no ground for assuming that it has any control over its instinctive behaviour.

The specialization of structure and of instinctive behaviour, in accordance with a definite sequence of life-conditions, is even more remarkable in another of the many parasites which *Chalicodoma* unwittingly labours to nourish.   This time it is a fly (*Argyromœba*), which lays a minute egg on the outside of the cell.   From this egg is hatched a slender threadlike

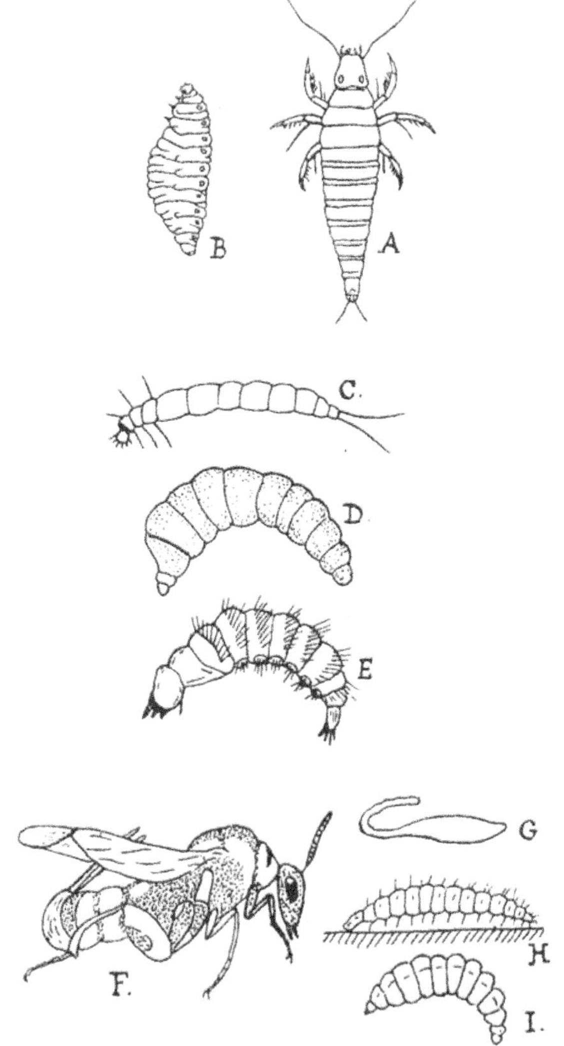

Fig. 13.—Insect Larvæ. A, B, of *Sitaris*; C, D, E, of *Argyromœba*; G, H, I, of *Leucopsis*; F, imago of *Leucopsis* (after Fabre).

worm, barely one-twentieth of an inch in length (Fig. 13, c).
It has three pairs of longish bristles near the anterior end, and
a single yet longer pair at the hinder extremity. These aid
it in creeping over the wall of the cell. Its small head is
armed with short, stiff bristles. For many days it wanders
over the surface of the cell, inserting its bristly head into each
minute cranny and crack. Throughout this long period it
has never a bite nor sup. Probably many of them never
succeed in finding a crevice by which they can effect an
entrance, but those that do manage to wriggle in undergo a
change, lose their bristles, and develop a minute suctorial
mouth, through which the contents of the larva are absorbed
into their swelling bodies (Fig. 13, d). When fully grown
they are quite helpless, and unable to get out from the cell in
which they are now imprisoned. For months they lie quiescent,
but in the succeeding spring they pass into a pupal condition
very different from that of most flies. The relatively large
head is armed with strong spines; the middle region bears
bristles directed backwards; the posterior end has short spines
(Fig. 13, E). Fixing itself to the interior of the cell by the
latter, it strikes with its armoured head repeated blows on
the walls of its prison until a breach is at last made, and
sufficiently enlarged to form a suitable exit. Then the pupa-
skin bursts, and the imago insect emerges and flies off. At
each stage of life there is the closest relation between structure
and behaviour, and each is equally adapted to a biological end
of which the creature has never had an opportunity of gaining
any experience.

Exceedingly multifarious are the ways in which insects
thus provide for the future of young they will never see.
Antherophagus lives in flowers, and is believed to seize with its
mandibles humble bees, which then unwittingly bear the
parasitic beetle to the nests in which alone the larvæ have
been found. The larvæ of our common oil-beetle (*Meloë*) are
parasitic on the bee, *Anthophora*. It deposits its ten thousand
eggs without observable discrimination; but the active young
larva instinctively seizes and attaches itself to any hairy object.

Thousands must go astray. They have been found on hairy beetles, flies, and bees of the wrong genus. Some, however, become thus attached to the one suitable species, and are conveyed by the *Anthophora* to her nest, where they promptly eat the egg she lays. It is not difficult to picture to one's self how this incompletely evolved instinct might be further perfected by natural selection, through the survival of those females which laid their eggs in the haunts of the bee-host. And such an advance in instinctive behaviour is seen in another and rarer beetle—*Sitaris*. Her eggs are laid in August near the entrance to a nest of the *Anthophora*. In September they hatch to form larvæ, which hibernate in groups till the following spring. Then they become active (Fig. 13, A), and attach themselves to hairy objects. Being near the *Anthophora* nest, there is an increased chance of their fastening upon this bee. The chance is still far from good, for if this were so, we should not find that the *Sitaris* laid as many as two thousand eggs. Still, on these grounds, we may presume that its chance of survival is about five times as good as that of *Meloë*, which lays ten thousand eggs. The larva is said generally to attach itself to a male bee, which is hatched earlier than his mate, and to pass on to the female at the nuptial period ; but in any case it eventually slips on to the egg that she lays. This forms the food of the larva during the remainder of this stage of its existence. It then moults and assumes a new form, capable of feeding on the honey (Fig. 13, B) ; and, after further changes, becomes a pupa, and then assumes the imago condition.

In these cases the advantage is wholly on the side of the parasite. But there are cases of close relationship between insects and flowering plants where the instinctive behaviour gives rise to reciprocal benefit. The Yucca is a genus of American Liliaceous plants, with large pale sweet-smelling flowers ; and these are dependent for fertilization on the instinctive behaviour of a small straw-coloured moth of the genus *Pronuba*. Just when the Yucca plant blossoms in the summer, the moths emerge from their chrysalis cases. They mate ; and the female then flies to a flower, collects a pellet of pollen from the anthers, proceeds to another flower, pierces the pistil with her sharp ovipositor, lays her eggs among the ovules, and finally darting to the stigma stuffs the pollen pellet into its funnel-shaped extremity (Fig. 14).   If the flower be not thus fertilized the ovules do not develop ; and if the ovules do not develop the grubs which are hatched from the moth's eggs die of starvation. There are enough ovules to supply food to the grubs, and leave a balance to continue the race of Yuccas.

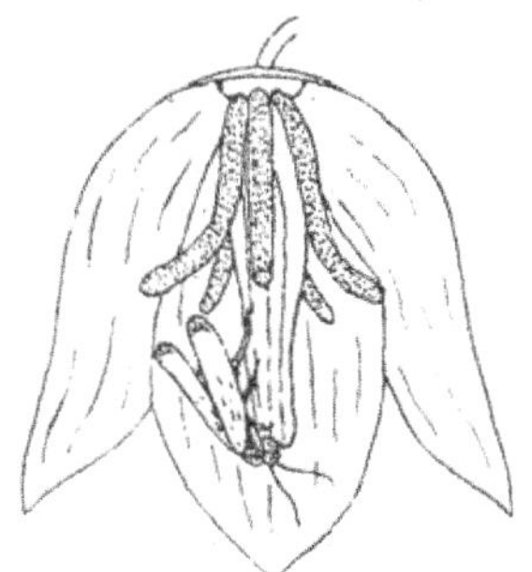

Fig. 14.—Yucca Flower and Moth.

Whether the female moth is attracted to the flower by sight or smell, we do not know.   And whether the male finds the female, in the case of the Yucca moth, through scent, we are not in a position to state with certainty.   It has, however, been shown that in certain moths * some odour emitted by the female is the attractive stimulus, affecting sense-organs situated on the antennæ of the male.   To females confined in an opaque vessel over the mouth of which gauze was tied, the males came in numbers ; but when a clear glass vessel was inverted, and sand was packed round the mouth, so as to prevent the escape of air from the interior, no males came, though the imprisoned females were clearly visible.   If the antennæ of the males were either removed or coated with shellac the moths failed to notice the females even when close to them.   In what way the intact male is made aware of the direction from which the scent comes, we do not know—possibly by differential stimulation in the antennæ, the moth instinctively turning in the direction of greater stimulation.   It will be seen, therefore, that in the case of the behaviour of the Yucca moth— behaviour which is essential to the biological end of repro-

* See A. G. Mayer " On the Mating Instinct of Moths." *Ann. and Mag. of Nat. Hist.*, ser. 7, vol. v., Feb., 1900, p. 183.

duction—there is much detail concerning which we are ignorant. But for our present purpose the important point to notice is that the procedure of the female cannot be due to imitation ; nor can it be the outcome of individually acquired experience ; for the method of procedure is not gradually learnt, but is carried out without apparent hesitation the first and only time the appropriate occasion presents itself. Not only does the moth take no heed of her grubs, but they are so placed that she could not in any case ascertain by observation that only if the ovules are fertilized do her offspring thrive. She cannot possibly know what effect the stuffing of the pollen on to the stigma exercises, or indeed whether it have any effect at all. And yet generation after generation these moths collect the pollen from the anthers and bear it to the stigma. Spence's words "without knowledge of the end in view" are amply justified in this case, as in other cases of typically instinctive behaviour.

III.—The Instinctive Behaviour of Young Birds

Since it is easy to hatch birds of many species in an incubator, and to rear them under conditions which not only afford facilities for observation but exclude parental in-fluence, their study has special advantages. One can with some approach to accuracy distinguish the instinctive from the acquired factors in their behaviour.[*]

The callow young of such birds as pigeons, jays, and thrushes are hatched in a helpless condition, and require constant and assiduous ministration to their elementary organic needs. Most of their instincts are of the deferred type. But pheasants, plovers, moor-hens, domestic chicks, and ducklings, with many others, are active soon after birth, and exhibit powers of complex co-ordination, with little or no practice of the necessary limb-movements. They walk

---

[*] Some of the observations on which the summary of results given in this section are founded are presented in some detail in "Habit and Instinct," pp. 29–100.

and balance the body so soon and so well as to show us that this mode of procedure is congenital, and has not to be gradually acquired through the guidance of experience. Young water-birds swim with neat and orderly strokes the very first time they are gently placed in water. Even little chicks a day or two old can swim well. Dr. Thorndike, who draws attention to this fact,* appears to accept the view,

FIG. 15.—Newly-hatched Chick swimming.

suggested by Dr. Bashford Dean, that the movements are not those of swimming but only of running. I have carefully watched the action through the glass walls of a tank and compared it with that of a young moor-hen. In the two cases it is quite similar in type, and the type appears to be different from that of running, though it is perhaps hard to distinguish

* *Psychological Review*, May, 1899, p. 286.

the two. In any case, the hand over hand action is well co-ordinated, and is very different from a mere excited struggle. Chicks twenty-six hours old taken straight from the incubator drawer, before they had taken food, made directly for the side of the tank and tried to scramble out. They gradually sank deeper through the wetting of the down, but could keep afloat for from two to three minutes. I have made observations on chicks of various ages from twenty-four hours to a month, and find in all cases similar results ; but with the older birds the flapping of the wings and more vigorous action cause them to get water-logged more rapidly. There is some apparent distress with cries ; but less than one might expect under the circumstances. For the purposes of the above illustration Mr. Charles Whymper had before him a sketch I made of the leg-action, and instantaneous photographs of the chicks swimming for which I am indebted to my colleague Mr. George Brebner. I have not observed the behaviour of an adult hen when placed in the water. Dr. Thorndike says, "there is no vigorous instinct to strike out toward the shore," she "will float about aimlessly for awhile and only very slowly reach the shore." But Mrs. Foster Wood informs me that she has seen a hen leap into a pond after her brood of ducklings and swim to the other side, a distance of twenty feet.

Diving, in water-birds, is also an instinctive mode of behaviour ; and this is obviously a more difficult procedure than swimming, one further removed from reflex action. And careful observations have placed beyond question the fact that flight is also instinctive. A swallow, for example, taken from the nest under conditions which made it practically certain that it had never yet taken wing, exhibited guided flight, and attempted to alight on a suitable ledge. Of course flight is generally a deferred instinct, and is not performed until the wings have reached a suitable state of development. An instinctive response, which may perhaps be regarded as one of its initial stages, is seen in quite young chicks. If placed in a basket, and rapidly lowered therein through a foot

or two, the chick will extend its skinny and scarcely feathered
wings. But though, from the usual conditions of development,
flight in birds is a deferred instinct, yet in exceptional cases
it may be connate. The mound-builders (*Megapodes*) of the
Australian region are hatched from large eggs in warm earth

Fig. 16.—Nestling Megapode, to show the well-developed wings. (From
Dr. R. Bowdler Sharpe's " Wonders of the Bird World.")

or sand, and are not tended by the parents. So well fledged
are these birds that they can fly the day they emerge from the
egg. Dr. Worcester, while digging in one of their mounds,
made an unsuccessful attempt to seize one which was newly
hatched ; but it flew several rods into thick brush, and this

notwithstanding the fact that it had probably never before seen the light of day.

It must not be supposed that, in adducing flight as an example of instinctive behaviour in birds, we are contending that it is this and nothing more throughout life. The inference to be drawn from the facts of observation is rather that instinct provides a general ground plan of behaviour which intelligent acquisition, by enforcing here and checking there, perfects and guides to finer issues. Few would contend that the consummate skill evinced in fully developed flight at its best, the hurtling swoop of the falcon, the hovering of the kestrel, the wheeling of swifts in the summer air, the rapid dart and sudden poise of the humming bird, the easy sweep of the sea-gull, the downward glide of the stork—that these are, in all their exquisite perfection, instinctive. A rough but sufficient outline of action is hereditary; but the manifold graces and delicacies of perfected flight are due to intelligent skill begotten of practice and experience.

There are many little idiosyncracies and special traits of flight which are probably instinctive—such as enable an ornithologist or a sportsman to recognize a flying bird from a distance. And the same is true of other modes of behaviour. The observer of young birds cannot fail to note and to be impressed by many of these. The way in which a little moor-hen uses its wings in scrambling up any rough surface is very characteristic; so, too, is the manner in which a guinea-chick runs backwards and then sideways at a right angle when one attempts to catch him. If suddenly startled, moor-hens and chicks scatter and hide; plovers drop and crouch with their chins on the ground; pheasants stand motionless and silent. Knowledge of the ways of birds enables one to predict with tolerable accuracy how each kind will behave under given circumstances. That the actions are always precisely alike cannot be said with truth; but that the behaviour is so relatively definite as to be readily recognizable can be confidently asserted. That a moor-hen will flick its tail, that a chick will dust itself in the sand, that pheasants and partridges will scratch the ground, that a jay will go through certain actions in the bath, that the preening of the down will be carried out in particular ways—moor-hens, for example, wringing out the water in a peculiar manner,—and that all these, and many other modes of behaviour, will be presented in relatively definite ways : all these are, to borrow the phrase of Dr. Peckham's, so characteristic of the several groups of birds, that they would be an important part of any definition based upon behaviour. And there can be no question that they are instinctive. They may indeed seem trivial and commonplace, scarcely worthy of special note ; but they serve to show in how many details organic heredity lays the foundation for future behaviour, and affords groups of data for effective consciousness to utilize.

To show the instinctive nature of such behaviour, the following examples will suffice. One of a batch of moor-hen chicks showed once, and once only, when a week old, an incipient tendency to bathe in the shallow tin of water which was placed in their run, but soon desisted ; nor was the action repeated, though he and the others enjoyed standing in the water. Five weeks later one of the batch was taken to a beck. He walked quietly through the comparatively still water near the edge ; but when he reached the part of the stream where it ran swiftly and broke over the pebbles, he stopped, ducked, and took an elaborate bath, dipping his head well under, flicking the water over his back, ruffling his feathers, and behaving in a most characteristic manner. Each day thereafter he did the same, with a vigour that increased up to the third morning, and then remained constant. The same bird some weeks later was swimming in a narrow part of the stream, with steep banks on either side, when he was frightened by a rough-haired pup. Down he dived, for the first time in his life ; and after a few seconds his head was seen to appear, just peeping above the water beneath the bank.

Ten days after receiving two nestling jays I placed in their cage a shallow tin of water. They took no notice of it, having never seen water before ; for they were fed chiefly on sopped food. Presently one of them hopped into it, whether attracted by the water or by accident it is difficult to say, squatted in it bending his legs, and at once fluttered his feathers, as such birds do when they bathe, though his breast scarcely touched the water. The other seized the tin in his bill, and then pecked at the water, thus wetting his beak. He, too, fluttered his feathers in a similar fashion, though he was outside the tin and not in the water at all. A little later the first again entered the tin, and dipped his breast well into the water ; this was followed by much fluttering and splashing. The bird took a good bath, as did the other shortly afterwards, and then spent half an hour in a thorough grooming, with much fluttering of the wings, the crest feathers being constantly raised and lowered, expressive of an emotional state.

Now, in these cases it would be impossible to say whether the behaviour was carried out in the manner characteristic of the species, prior to experience and independent of imitation, on the basis of mere casual and chance observation. But in these cases the whole life-history of the individuals concerned was known ; and it can be asserted with confidence that the behaviour was hereditary, and not acquired by any gradual process of learning. Moreover, in each case there seemed to be such evidence as observation can afford, that internal emotional factors co-operated with the direct external stimuli in determining the nature of the behaviour. Whether such actions so far contribute to the well-being of the individual as to be of decisive advantage it is difficult to say. Some would contend that bathing is practised by birds merely for the pleasure it seemingly affords ; others would urge that it is a means of getting rid of troublesome and presumably hurtful parasites, to the attacks of which birds are peculiarly subject.

One of the most remarkable instincts of young birds is that of the cuckoo, which ejects eggs and nestlings from the home of its foster-parent. Mrs. Hugh Blackburn found a nest which contained two meadow-pipits' eggs, besides that of a cuckoo. On a later visit " the pipits were found to be hatched, but not the cuckoo. At the next visit, which was

after an interval of forty-eight hours, we found the young
cuckoo alone in the nest, and both the young pipits lying
down the bank, about ten inches from the margin of the nest,
but quite lively after being warmed in the hand. They were
replaced in the nest beside the cuckoo, which struggled about

FIG. 17.—Young Cuckoo ejecting nestling Meadow Pipit. (From Mrs.
Hugh Blackburn's sketch in " Birds from Moidart.")

until it got its back under one of them, when it climbed backwards directly up the open side of the nest, and hitched the
pipit from its back on to the edge. It then stood quite upright
on its legs, which were straddled wide apart, with the claws
firmly fixed halfway down the inside of the nest, among the
interlacing fibres of which the nest was woven, and, stretching
its wings apart and backwards, it elbowed the pipit fairly over

the margin, so far that its struggles took it down the bank instead of back into the nest. As it was getting late, and the cuckoo did not immediately set to work on the other nestling, I replaced the ejected one and went home. On returning next day, both nestlings were found dead and cold, out of the nest." * Here we have a definite account by an eye-witness, who sketched the young cuckoo, which was naked, blind, and could scarcely hold up its head. And her account, itself confirmatory of that given by Jenner in 1778, is confirmed by that of Dr. John Hancock,† who witnessed the ejection of a fledgling hedge-sparrow, which " was put over the edge of the nest exactly as illustrated by Mrs. Blackburn." The procedure is unquestionably instinctive.

The sounds uttered by young birds are sufficiently definite to be readily recognized and are susceptible of classification. In domestic chicks at least six notes may be distinguished. First the gentle " peeping " note, expressive of contentment. A further low note, a double sound, seems to indicate extreme satisfaction and pleasure. Very characteristic and distinct is the danger-note—a sound difficult to describe, but readily recognized. If a humble-bee, a black-beetle, a big worm, a lump of sugar—anything strange and largish—be thrown to the chicks, this danger-note is at once heard ; and it serves to place others on the alert, though this is perhaps the outcome of experience. Then there is the cheeping sound, expressive apparently of a state of mild dissatisfaction with the present state of affairs. It generally ceases when one throws some grain, or even stands near them. Extreme dissatisfaction is marked by a sharper, shriller squeak, when one seizes them against their inclination. Lastly, there is the shrill cry of greater distress, when, for example, their swimming powers are subjected to critical examination. With pheasants a gentle, " peeping " note of contentment, a shriller cry of distress, and a danger-note, generically like, but specifically distinct from,

* " Birds from Moidart and Elsewhere," p. 107. Edinburgh : Douglas.
† *Transactions of Northumberland and Durham Natural History Society,* vol. viii., p. 213.

that of the chick, are early differentiated. The complaining note of the partridge is uttered six or seven times in quick succession, followed by a pause. The note of the plover is high-pitched, and much like the familiar cry of the adult bird, to which it owes its popular name of "peewit." So, too, the guinea-fowl in down utters from the first notes quite characteristic of its kind, while its danger-note is not unlike that of the chick or pheasant. The piping of ducklings is comparatively monotonous, and there does not seem to be a definite danger-note. With moor-hen chicks, even on the first day, two notes are well-marked—a call-note, lower in pitch than that of the chick, and rather harsh and raucous, and a " tweet, tweet " of pleasure, something like the contented note of a canary. Later, five or six notes are differentiated, the most characteristic of which is the harsh "crek, crek," when the little bird is from any cause excited. It is uttered in a crouching attitude, with head thrown back and wings held outwards and forwards, waving to and fro in a very characteristic manner. That this has suggestive value for other moor-hen chicks is shown by its distinctly infectious effect ; if one bird has cause to utter the note and strike the attitude others follow suit. While clearly instinctive in their mode of occurrence, while they seem to show well the co-operation of an internal emotional factor, their biological value seems to lie in their suggestive effect on other members of the brood. They form an elementary but sufficient social bond.

If these notes afford evidence of an incipient social factor, the instinct of pecking is distinctively individualistic. Chicks peck with considerable but not complete accuracy of aim at practically anything of suitable size at suitable distance ; but it is through experience that they learn what to select for food and what to reject or leave untouched. Moving objects, however, are more readily pecked at than those which are still ; and the instinctive response seems to be stimulated if one tap on the ground near the object, or move it with a pencil, thus simulating the action of the hen. And this is even more marked with pheasants and partridges. Plovers seize small worms with an avidity which looks like an inherited recognition of natural food. Pheasants and partridges also appear to be specially affected by worms, and when one of them seizes a worm for the first time, he shakes it and dashes it against the ground. Chicks, a week or ten days old, also seize a largish fly or bee with a dash, and maul it on the ground, throwing it on one side before again approaching it. And such birds seem to show an instinctive tendency to bolt with such treasures as caterpillars or small worms. Moor-hens cannot at first be induced to take food from the ground. It has to be held above them, whereupon they crouch down, with head and neck held back, opening their beaks more like the callow young of nursling birds ; but they also strike upwards at the object—these modes of behaviour being, no doubt, correlated with the manner in which the mother moor-hen normally feeds her young from her beak during the early days of life. Callow fledglings, such as young jays, simply open their mouths, gaping widely to be fed. And many will respond in this way to such a note as a low whistle, as may readily be seen with swallows. But at a later age such birds show instinctive modes of reaction of a more complex type. A jay, for example, was offered a summer chafer or June bug, seized it at once in his bill, and tried to place his foot on it. Then he hopped down on to the floor of his cage, dropped the beetle, seized it again as it crawled off, and after two or three attempts swallowed it, tossing it back from the point of his bill into the throat. This was the first time he took food from the ground or swallowed it in this manner.

On the whole, there seems to be much inherited definiteness of co-ordination, and some tendency to respond in a definite manner to specific stimuli. That there should not be more differentiation in this respect than observation discloses is probably due to the fact that the parent birds afford, under natural conditions, so much guidance in the selection of food. Since the solitary wasp unerringly seizes its appropriate food, since it responds instinctively to specific stimuli, there would seem no reason why birds should not show similar instinctive differentiation. But one must remember that in the case of the wasp there is no parental guidance; the insect is more completely dependent on instinct than is the bird to whose needs the hen assiduously ministers.

It is at first sight surprising that such birds as chicks and pheasants do not peck instinctively at still water. When a shallow vessel containing water was placed among some little chicks, several of them ran repeatedly through the water, but took no heed of it. Then, after about an hour, one of them standing in the vessel pecked at his toes, and at once lifted his head and drank freely with characteristic action. Another subsequently pecked at a bubble near the edge, and then he too drank. In fact, the best way of inducing them to drink is to scatter some grains of food in the tin; they peck at the grains, which catch their eye, and incidentally find the water, and the touch of water in the bill at once leads to the characteristic response and congenitally definite behaviour. That the sight of a still surface does not itself suffice to evoke this behaviour is probably again due to the fact that under nature the hen guides them and pecks at the water, when they follow her bill.

One fact which must be constantly borne in mind is that what is inherited is instinctive co-ordination, often related to a definite stimulus, not instinctive knowledge. A chick pecks at a grain when it is at a suitable distance, not because instinct provides him with the knowledge that this is something to be seized and tested, but because he cannot help doing so. He is so organized that this stimulus produces that result through an organic co-ordination that is independent of conscious knowledge or experience. How definite is the inherited co-ordination is shown by many observations. A young pheasant, only a few hours old, was taken from the incubator drawer, and held snugly while a piece of egg-yolk was moved before his eyes with the aid of fine forceps. He did not peck at it, but followed with movements of his head every motion of the object in a narrow circle. Simple as this action seems, it presents a striking example of co-ordinated movements accurately related to movements in the visual field, the whole performed without any opportunity for learning or practice, and less than half an hour after the bird was taken from the drawer of the incubator and first saw daylight. Psychologists sometimes puzzle their heads over the question how and by what steps the field of vision and the field of movement are brought into relation with each other; but in such a case as this, the problem ceases to be primarily psychological. The relation is purely organic ; the conscious data are grouped from the outset. With young jays there was no such co-ordination at first ; and when they began after a few days (about twelve or fourteen) to follow an object with the head and eye, the movements were at first jerky. But a week later, when I swept the food through a circle a foot in diameter in front of their cage, it was followed smoothly and evenly. Here a certain amount of learning and practice, absent in the case of the pheasant, was required. And it is difficult to say what proportion of the final result was acquired, what proportion hereditary ; but probably the behaviour is in the main instinctive, though somewhat deferred.

One more example, perhaps even more trivial in the eyes of some people, may be given. A duckling a few hours old will scratch the side of his head. It is true he may topple over in the process, through insufficient powers of balance, for the simultaneous performance of poising on one leg and having a good scratch with the other is no easy matter. But let not either our familiarity with such behaviour, nor some observed and laughable failure on the part of the duckling, blind us to the fact that this is a congenital activity, and one of no little complexity, indicating a definite organic nexus. A local irritation sets agoing movements of the hind limb of that side through which just that particular spot is scratched in the absence of any previous practice, any learning to localize the spot. There can be no question that such inherited co-ordinations, whether perfect from the first, or with deferred perfection and some aid from acquisition, afford ready-made data to consciousness, which are of the utmost service in the guidance of subsequent behaviour. The two-days-old chick, with the aid of this instinctive co-ordination, performs well a number of actions, which, had she to consciously learn them all, would probably be still but half mastered when she was a skinny old hen.

Our whole treatment of instinctive behaviour has been based on the assumption, already to some extent justified, that experience is not inherited. If it be hereditary, how comes it that chicks show no recognition of still water, which must have been familiar to the experience of generation after generation of birds ? How comes it that they do not even seem to recognize their natural parent and protector, the hen ? Two chicks ten days old were taken to the yard whence were derived the eggs from which they were hatched, and were placed about two yards from a hen which was clucking to her brood. They were not in a frightened condition, for they stood on my hand and ate grain from it, scratching at the palm. But of the clucking of the hen they took no notice whatever. The same results were obtained with other chicks thirteen days old. Was this due, as Spalding suggested, to loss of the instinctive response which was perhaps present at an earlier age? Seemingly not. For a chick was taken at the age of two and a half days to its own mother, which had three chicks. These followed her about, and ran at once to her when she clucked and pecked on the ground. The little stranger, however, took no notice, nor did he show any tendency either to go to the hen or to follow the three chicks, having been purposely brought up alone. When the hen took her little brood under her wing, the stranger was placed close to her. She clucked, and seemed anxious to entice and welcome the little fellow, seizing an oat-husk and dropping it before him ; but he remained indifferent, walking away and standing in the sunshine. After about forty minutes he seemed more inclined to go with the other chicks, but still ignored the existence of the hen. The natural instinctive tendency seems to be from the first to nestle under anything ; and there is the hen provided by nature for the purpose. By experience the chicks grow accustomed to her fussy ways, as they grow accustomed to the ways of such a foster-parent as the writer of these pages. Still, though there is, apparently, no instinctive knowledge of the hen as their natural protector, and though I have seen no observable response to the clucking sound, this must not be taken as necessarily implying that there is no instinctive response to any of her modes of behaviour. There is such a response to her pecking on the ground ; there is probably such a response to her danger-note ; and there may be many other such instinctive modes of behaviour related to her actions. How far they extend can only be ascertained by patient observation ; and such responsive behaviour need not imply any instinctive knowledge begotten of inherited experience.

We may now summarize some of the general conclusions which may be drawn from observations of instinctive behaviour in young birds.

1. That which is inherited is essentially a motor response or train of such responses. Mr. Herbert Spencer's description of instinct as compound reflex action is thus justified.

2. These often show very accurate and nicely-adjusted hereditary co-ordinations.

3. They are evoked by stimuli, the general type of which is fairly definite, and may in some cases be in response to particular objects.

4. They are also generally shown under conditions which lead us to infer the presence of an internal factor, emotional or other.

5. There does not seem to be any evidence of inherited knowledge or experience.

IV.—The Conscious Aspect of Instinctive Behaviour

In our definition of instinctive behaviour all positive reference to the presence of conscious states was omitted. By some writers, however, the fact that it is accompanied by conscious-ness is regarded as a distinguishing feature of instinct. Romanes introduced his definition with the words : * " Instinct is reflex action into which there is imported the element of consciousness." And he emphasized the conscious aspect when he said : " The term comprises all those *faculties of mind* which are concerned with conscious and adaptive action, antecedent to individual experience." Professor Wundt also lays some stress on the conscious accompaniments of instinctive activities which, he says,† " differ from the reflexes proper in this, that they are accompanied by emotions in the mind, and that their performance is regulated by these emotions." The definitions of other writers express or imply the presence of consciousness in differing modes and degrees, culminating in the hypothesis of inherited knowledge. Douglas Spalding, for example, said ‡ that "animals can forget the instinctive knowledge which they never learned ! "

Now, the exclusion from our definition of direct reference to the conscious aspect must not be taken to imply that instinctive behaviour is a mere matter of unconscious automatism ; nor even that it is unprofitable to discuss how much consciousness there may be, of what sort, and how distributed. All that it does imply is, that the amount, nature, and distribution of consciousness cannot well be introduced into a definition the object of which is to help us to distinguish certain observable types of behaviour from others. In a word, the definition given is biological and objective, and is to be accepted or rejected without prejudice to such psychological considerations as those upon which we have now briefly to enter.

The first thing we have to decide is how much we are to include, from the psychological standpoint, under instinct. For we may take either a broader or a narrower view of the matter ; and which of these we adopt will make much difference in our conclusions. Let us first deal with the narrower. We have said above that what is hereditary in instinctive behaviour is the co-ordination. Now, such co-ordination of

---

* " Mental Evolution in Animals," p. 159.

† " Lectures on Human and Animal Psychology," p. 401.

‡ " Instinct and Acquisition," *Nature*, vol. xii., p. 507.

movements into a finished and appropriate act is due to a nicely graded distribution of efferent nerve-waves to the several muscles concerned, so that these muscles may be caused to contract in due order, and each to just the right extent. But efferent nerve-waves as such, and their mode of distribution by the nerve-centres, are in all probability unconscious, while the contraction of the muscles is a purely organic matter. If, therefore, we narrow our conception of instinct so as to include only the co-ordinated act by itself, excluding all reference both to the stimuli which are its antecedents, and to the effects in consciousness which its performance may produce ; and if the data for consciousness are in all cases supplied through afferent channels ; then there seems to be no escape from the conclusion that instinctive behaviour as such may be, and probably is, altogether outside the individual consciousness. It should be noted, however, that on this view it is the co-ordination in itself that is, if not unconscious, at any rate independent of the stream of experience.

Now, in the first place it is convenient so far to broaden our conception as to include under the head of instinctive behaviour, in its conscious aspect, not only the co-ordinated act but the data which its performance affords to consciousness. It may indeed seem that we are here trying to draw a distinction where no real difference exists. The physiological distinction is, however, not only clear and undeniable, but quite easily understood. For the sake of illustration let us take the case of an intentional action, such as glancing up from the words we are reading to the clock. Efferent waves course along several motor nerves to the six muscles by which each eye is moved, and to the muscles of accommodation within the eye. These muscles are called into duly co-ordinated activity, by which our vision is focussed upon the clock-face. This is one part of the physiological procedure—that by which the intended result is attained. But there is a second part readily distinguishable from the former. As the eyes move, afferent messages course inwards from the muscles or the eye-sockets and their neighbourhood ; and it is these incoming waves which

afford data to consciousness, telling us that the movements are
in progress or have been effected. The nerves involved in the
latter part are quite different from those concerned in the
former part, and they proceed to areas of the brain differently
situated from those whence the efferent waves issued. Thus
it is in all cases of movement ; the efferent nerves call the
muscles into play ; the afferent nerves bring information that
the movements are carried out. It is through the latter that
data are unquestionably afforded to consciousness.

But in the case of any complex action—and, as we have
seen, instinctive behaviour is often remarkably complex—the
information that the action has begun comes in before the
behaviour is completed. Practically we may say that any
given stage of performance and the consciousness it evokes are
simultaneous ; for though in strictness the one lags just a
little behind the other, yet they are so nearly coincident in
time that we may disregard the interval between them. Such
being the case, therefore, we may fairly regard instinctive
behaviour as capable of introducing important elements into
the conscious situation.

But not only does instinctive behaviour thus introduce
important elements into the conscious situation, it is also
called forth by stimuli which themselves afford not less
important elements. To exclude these from any consideration
of instinct, in its conscious aspect, would render the treatment
of the problem so incomplete as to be wholly unsatisfactory
from a psychological point of view. Can we believe that
when the moor-hen dived, as it never had dived before, at the
sight of the rough-haired pup, the vivid experience of that
strange and disquieting intruder did not enter into, and form
a prominent feature in, the conscious situation ? If we are to
consider the conscious aspect at all, we must try and grasp the
situation as a whole. And on these grounds we may yet
further broaden our conception so as to include, from the
psychological point of view, not only the behaviour itself, and
its effects in consciousness, but also the conditions under which
the complex actions are called into play. If, then, we accept

this position, and agree to use the term "instinct" for our present purpose in a comprehensive sense, we may now proceed to consider very briefly the nature of the elements which enter into the instinctive situation.

First, there are the external stimuli affecting one or more of the sense organs, and thus evoking consciousness ; and secondly, there are internal factors, having their source in the condition of the body, or its parts and organs. It is convenient to take these two together, so that we may see what relationship they bear to each other. Both seem to be present, and to co-operate in a great number of instinctive acts. In the behaviour connected with feeding, for example, an internal element of hunger co-operates with the external presentation of the appropriate food or prey. So, too, with the instincts concerned in the propagation of the race. Looking at the matter generally, we may regard the internal factors of the kind with which we are now dealing, as giving rise to a want or need, passing in some cases into a state of craving. In themselves such conscious states are in their inception exceedingly indefinite ; for a want can only be rendered definite in experience by its appropriate satisfaction. In many cases of instinctive behaviour the indefinite want and the particular and duly related stimulus seem to lead, without prevision and by a blind impulse, to the performance of those acts which will afford the unforeseen satisfaction. And when once this satisfaction has been attained, subsequent wants or needs of like character will no longer be indefinite ; nor will future behaviour of the same kind be thereafter wholly instinctive, for it can never again be prior to, or independent of, experience.

Granted, however, that a felt need of some kind, indefinite at first but none the less real, is present in many cases as a spur to instinctive behaviour ; is it in all cases a necessary factor ? May we say that this distinguishes instinctive from merely reflex action ? The question is, from the nature of the case, exceedingly difficult to answer. But without going so far as to say that reflex action may be unerringly distinguished from instinctive behaviour by the absence of any such internal factors, we may perhaps, at any rate, go so far as to give provisional acceptance to the view that in instinct these wants and felt needs enter into the conscious situation in a manner and to a degree that are so far distinctive—which seems to be the position adopted by Professor Wundt.

There is, however, a further relation between the external stimulus and these internal factors which is presumably of no little importance. The stimulus intensifies the want, or may in some cases call it into existence. Just as a whiff from the kitchen may lead us to realize that we need the satisfaction that will erelong be presented at table, so may the sight of his mate in the spring evoke in the breast of the yearling sparrow a need, having its source in morphological and physiological changes, that spurs him on to the courtship that shall lead to its due satisfaction. Popular attention has, indeed, been so naturally drawn to the internal needs or wants with which we are now dealing, as to give them an almost exclusive monopoly of the term "instinct," which thus often comes to be regarded as a connecting link between the stimulus and the act. The sight of a mouse, for example, is said to call forth the instinct of the cat, which is satisfied by her pouncing upon it. And so it comes about that, while the biologist fixes upon the instinctive act as the essential feature, the psychologist is apt to regard the impulse * which prompts to action as the more central and characteristic element. We are here endeavouring to combine both these points of view.

To come to closer quarters with the relationship which holds good between the external and internal elements, it appears that, when the stimulus evokes or intensifies the want or need, this is probably effected by efferent waves which call the organs or parts into tonic action, of which the animal becomes conscious through the afferent messages which come in from them to the sensory centres ; in much the same way as the whiff from the kitchen takes effect on the salivary and other glands, and throws the organs of digestion into a felt preparedness for the fulfilment of their functions. But it

* On the nature of impulse, see *infra*, p. 235.

may have other and more indirect consequences. When the moor-hen dived to escape from the obtrusive puppy, his heart-beat was probably affected ; he had, perhaps, an uncomfortable sinking in his gizzard ; his breathing was short and laboured ; and he experienced creepy sensations in the skin and around the feather-roots. Such we may suppose were the accompaniments or sources of the emotional state of fear or alarm. And they presumably entered with no little vividness into the conscious situation at the moment of instinctive action. In all those cases in which the behaviour is associated with such an emotional state as anger or fear, the external stimulus seems to produce widely-spread effects on the glands, respiratory organs, heart and blood-vessels, skin and other parts, as well as the more direct response in productive action. And all this must enter into the conscious situation, contributing largely, as we shall hereafter see, to the emotions in their instinctive origin.

Enough has now been said to indicate with sufficient clearness the kind of co-operation and mutual relationship which subsists between the external and the internal factors in the conscious situation which leads to instinctive behaviour. We have seen that, not improbably, some organic prompting is always present in greater or less degree. But the question still remains whether anything like a definite and particular external stimulus is in all cases a necessary factor.

When the predaceous larva of the water-beetle, *Dytiscus*, ceases to feed, and, creeping into the moist earth near the pond's edge, makes a hollow cell in which to enter upon its pupal sleep, there does not seem to be any well-defined stimulus from the outer world which can be said to initiate the behaviour of whose purport the larva can have no idea. Some inner need seems to impel the creature to this necessary but as yet unknown course of action ; and this appears to constitute, if not the sole, at least the preponderant element in the conscious situation. In healthy young birds and other animals there is after the rest of sleep a certain exhilaration and exuberance of spirits which seemingly leads to characteristic action ; dancing, flapping of the wings, running hither and thither in short quick spurts, and so forth. No doubt in such cases external stimuli are present, and contribute in some degree to the effects produced; but they do not seem to be particularized so that one can say that just this or that well-defined stimulus is necessary to give rise to the observed behaviour. In the case of migration, too, an internal factor—the nature of which we do not know—is probably as strong as if not stronger than any influence from without. While, therefore, we may say that some external factors are frequently, not improbably always, contributory, we must add that observation does not enable us in all cases to define them with any approach to accuracy; and, further, that promptings from within seem in some instinctive acts to be the most important elements in the conscious situation.

It now only remains to draw attention to the fact that the effects of the behaviour, as the animal becomes conscious of the performance of the acts concerned, serve to complete and render definite the conscious situation. Consciousness, however, probably receives information of the net results of the progress of behaviour, and not of the minute and separate details of muscular contraction. These net results, having thus entered presentatively into the situation, are subsequently susceptible of re-presentative recall, when the recurrence of certain salient elements serve to reproduce the essential features of the situation of which experience has been gained on a former occasion. Hence, as has already been noted, it is only the *first* performance of an instinctive action which can be described as prior to experience. The second time the deed is done it is done by an animal which has had opportunity of gaining experience on the foregoing occasion. And then it may be done with a difference, with some acquired modification of performance. By the repetition of the slightly modified behaviour the effects of habit are introduced, and thus acquired peculiarities of action are established as individual traits. We must not forget that, in a large number of cases, so-called instinctive behaviour, as presented to observation, has lost through modified repetition its original purity of type. The acts we see are often the joint products of heredity and individual acquisition, the inherited co-ordination having been supplemented or otherwise altered through experience.

Even in the case of the very first exhibition of such a deferred instinct as the moor-hen's dive, although that organized sequence of acts which constituted the behaviour as a whole had never before occurred, although there was no gradual learning how to dip beneath the surface, and to swim under water, still many of the constituent acts had been often repeated ; experience had already been gained of much of the detail then for the first time combined in an instinctive sequence. So that if we distinguish between instinct as congenital and habit as acquired, we must not lose sight of the fact that there is continual interaction, in a great number of cases, between instinct and habit, and that the first performance of a deferred instinct may be carried out in close and inextricable association with the habits which, at the period of life in question, have already been acquired. Instinct supplies an outline sketch of behaviour, to which experience adds colour and shading. Which predominates in the finished picture depends on the status of the animal. In the lower and less intelligent types the outline stands out clearly, there being but little shading to divert our attention from the clear firm lines inscribed by heredity ; but in the higher and more intelligent animals, the deft pencil of experience has added so much and has interwoven with the fainter outline so many new and skilfully introduced touches, that the original sketch is scarcely distinguishable unless we have carefully watched from the beginning the gradual development of the picture.

V.—The Evolution of Instinctive Behaviour

We have seen that Professor Wundt distinguishes two classes of instinctive acts : first, those which are *acquired* or have become wholly or partly mechanized in the course of individual life ; secondly, those which are *connate* or have been mechanized in the course of generic evolution. "The laws of practice," he says,* "suffice for the explanation of the acquired instincts. The occurrence of connate instincts renders a subsidiary hypothesis necessary. We must suppose that the physical changes which the nervous elements undergo can be transmitted from father to son. . . . The assumption of the inheritance of acquired dispositions or tendencies is inevitable if there is to be any continuity of evolution at all. We may be in doubt as to the extent of this inheritance ; we cannot question the fact itself."

Now, the application of the term " instinct," both to acquired and to connate behaviour, seems to prejudge the question of their genetic connection. And since we have the well-recognized term *habits* for actions the performance of which becomes automatic through frequency of repetition, we may substitute this term, or the phrase *habitual acts*, for the "acquired instincts" of Professor Wundt. Modifying, therefore, his statement in accordance with this usage, the fact which, he says, we cannot question is that acquired habits are inherited as congenital instincts. This opinion has long been held : G. H. Lewes regarded instinctive actions as transmitted habits from which the intelligence, through which they were originally acquired, had lapsed. Darwin believed that such inheritance was a factor in the evolution of instinctive behaviour. Romanes distinguished instincts due to this mode of origin as "secondary ; " reserving the term "primary" for those attributable to natural selection, and describing those in which both factors co-operate as " instincts of blended origin." The late Professor Eimer, of Tübingen, going further than either Darwin or Romanes, reverted almost entirely to what we may term the Lamarckian interpretation. "I describe as automatic actions," he says,† "those which, originally performed consciously and voluntarily, in consequence of frequent practice come to be performed unconsciously and involuntarily. . . . Such acquired automatic actions can be inherited. Instinct is inherited faculty, especially is inherited habit." In

---

* " Lectures on Human and Animal Psychology," p. 405.
† " Organic Evolution," pp. 223, 263, 258, 279, 276, 298.

his discussion of the subject Eimer makes no express allusion to primary instincts ; but he attributes to lapsed intelligence some of those which were classed by Romanes as primary, and his tendency is to refer all instincts to the same source. " Every bird," he says, " must, from the first time it hatches its eggs, draw the conclusion that young will also be produced from the eggs which it lays afterwards, and this experience must have been inherited as instinct." Why, in the first instance, it must draw the conclusion from observation if it inherit instinctive knowledge, is not made clear. But our present purpose is to indicate, not to criticize, Eimer's position. He claims that " the original progenitors of the cuckoo, when they began to lay their eggs in other nests, acted by reflection and design." Of the behaviour of mason wasps and their allies, which provide their young with paralyzed but living prey, he exclaims, " What a wonderful contrivance ! What calculation on the part of the animal must have been necessary to discover it ! " Of the instincts of neuter bees he remarks, " Selection cannot here have had much influence, since the workers do not reproduce. In order to make these favourable conditions constant, insight and reflection on the part of the animals, and the inheritance of these faculties were necessary." And he concludes, " Thus, according to the preceding considerations, automatic action may be described as habitual voluntary action ; instinct, as inherited habitual voluntary action, or the capacity for such action."

Turning now to the opposite end of the scale of opinion, we find that Professor Weismann, commenting on the supposed inheritance of acquired habit, says,* " I believe that this is an entirely erroneous view, and I hold that all instinct is entirely due to the operation of natural selection, and has its foundation, not upon inherited experiences, but upon variation of the germ." Ziegler and Groos in Germany, Whitman and Baldwin in America, Poulton and Wallace in England, either deny the existence of secondary instincts, due to the inheritance of acquired habits, or question the sufficiency of the evidence

* " Essays on Heredity " (1889), p. 91.

adduced in support of such transmission. In their explanation of the manner in which that inherited co-ordination, which is biologically the central fact in instinctive behaviour, has been evolved they rely mainly or entirely on the principle of natural selection.

What, then, were the facts which appeared to Romanes sufficient to justify a belief in the existence of a class of instincts dependent on inherited habit for their origin ? He tells us that he only gives a few examples "amongst almost any number" that he could quote. It is certainly unfortunate that, out of more than one hundred and fifty pages devoted to instinct in his work on "Mental Evolution in Animals," only three * are assigned to secondary instincts ; or six, if we include one dealing with inherited peculiarities of hand-writing in man, and two showing the force of heredity in the domain of instinct, "whether of the primary or secondary class." It is true that many pages are devoted to instincts of blended origin, but the co-operation of the Lamarckian factor is here rather assumed than proved. We must, however, be content to take the few examples that are actually given. They are four in number. First, that ponies in Norway are used without bridles, and are trained to obey the voice ; and that, as a consequence, a race-peculiarity has been established, for Andrew Knight says that it is impossible to give them what is called "a mouth." No details being given, this strikes one as rather thin as a matter of evidence. Secondly, Mr. Lawson Tait had a cat which was taught to beg for food like a terrier. All her kittens adopted the same habit under circumstances which precluded the possibility of imitation. Supposing the facts to be correctly reported, and granting that the owners of the kittens, presumably aware of the maternal propensity, did not take some pains to teach the offspring of such a parent to beg (and this does not present much difficulty), one can hardly found a scientific conclusion on so slight an anecdotal basis. Thirdly, instinctive fear is said to be an inherited acquisition ; which, fourthly, is lost by disuse. But, as we

* *Op. cit.*, pp. 196-198.

have already seen, modern investigation has placed this matter of so-called hereditary fear of natural enemies on a different footing. Pheasants, partridges, moor-hens, and wild duck show no fear of a quiet dog. If approached gently, in the absence of their parents, callow wild birds in their nest exhibit little alarm at the slow and gentle approach of man. Mr. Hudson's opinion has already been quoted, but will bear repetition; it is, "that fear of particular enemies is in nearly all cases the result of experience." And there is no evidence to show that, in those cases in which it is truly instinctive and not the result of experience, the instinctive behaviour is necessarily due to inherited habit and not to natural selection.

It cannot be said that the evidence for the supposed mode of origin of secondary instincts is sufficiently varied and cogent to carry conviction. On the other hand, there does seem some evidence which points to a different conclusion. When instinctive behaviour follows on a sensory impression, not only is the co-ordination hereditary, but there is an inherited linkage of stimulus and response. Thus in the solitary wasps the sight of the natural prey is followed by the appropriate modes of attack. The *Meloë* larva springs upon anything hairy. In chicks the sight of a small object at a certain distance initiates the act of pecking. In moor-hens and ducklings the stimulus of water produces the movements concerned in swimming. And so, too, in many other examples of instinctive behaviour, we infer from the observed facts that stimulus and response have an organic connection founded on hereditary links in the nervous system. Now, if such connection were due to inherited habit, we should expect them to be established wherever the experience to which they are related has been constant through many generations. How comes it, then, that the chick does not instinctively respond by appropriate behaviour to the sight of water? How comes it that young birds do not instinctively avoid bees, and wasps, and nauseous caterpillars? If the effects of ancestral experience be hereditary, one would surely expect that in these cases the connection between stimulus and response—a connection which passes into acquired habit—would have become congenital; that the habitual behaviour would have long ago become instinctive. But this does not appear to be the case. And with regard to disuse causing the loss of instinct, how comes it that young chicks swim with well-ordered leg-movements, though swimming is not an act that is habitually performed by the members of their race?

What, then, has the alternative hypothesis of natural selection to advance in explanation of these facts? On this hypothesis instinctive acts have biological value in such degree that they have become congenital through the preservation of adaptive variations. But if this be so, why does not the chick respond instinctively to the sight of that which is so essential to its existence as water to drink? In reply to this question it may be suggested that, under natural conditions, the hen teaches all her chickens to peck at the water, and thus shields them from the eliminating influence which gives rise to natural selection, in the absence of which the habit of drinking in response to the sight of water, though acquired by each succeeding generation of birds, has not become instinctive and congenital. Or, to put the matter from a slightly different point of view, the maternal instincts of the hen protect her chicks from any elimination in this respect; and in the absence of such elimination the habit has not been inherited as instinct. But though the hen can lead her young to peck at the water, she cannot teach them how to perform the complex movements of the mouth, throat, and head in actual drinking. In this matter, therefore, her own instinctive procedure does not shield them from the incidence of that elimination which leads to survival under natural selection. Those chicks which, on pecking the water, failed to respond to the stimulus by the complex behaviour involved in drinking would be eliminated, leaving those to survive in which the response had been congenitally established. Thus it would seem that, when natural selection is excluded, the habit has not become congenitally linked with a visual stimulus; but when natural selection is in operation, the response has been thus linked

with the stimulus of water in the bill. Whence we may infer
that the co-operation of natural selection is an essential factor
in the evolution of instinctive behaviour.

There are, however, cases of instinctive behaviour which
may seem too trivial and unimportant to be subject to the
sway of natural selection. There are numberless little idiosyncracies of behaviour which seem to be truly instinctive,
which are readily recognizable as distinctive traits, but which
can hardly be regarded as of sufficient biological value to
determine whether the creatures in which they are developed
should survive or be eliminated in the struggle for existence.
In many cases, however, these serve rather to distinguish the
detailed manner of behaviour than its biological end or
purpose. In different species natural selection may determine
the survival of those whose instinctive behaviour meets a
biological need. The relatively unimportant details, differing
slightly in each species, are mere adjuncts ; and since natural
selection deals with each species or inter-generating group
separately, the essential behaviour may in each case carry with
it the associated differences of manner. We must remember,
too, that, as in the matter of structure so in that of behaviour,
it is the animal as a whole that is selected for survival ; and
so long as the whole is adapted to the circumstances of life,
the associated differences of form or manner may share in,
without doing much to determine, survival. In any case these
little instinctive traits, if they are so trivial as to seem of small
value from the biological point of view, appear to be too
unimportant to have been intelligently acquired as habits.

Let us now consider one or two cases of instinctive behaviour
which would fall under Romanes's category of instincts of
blended origin partly due to natural selection, partly to the
inheritance of acquired habit. It is the custom of the house
martin to build beneath the eaves. Forsaking the ancestral
rocky haunts, it has been led to utilize the houses that
man has built. This has all the appearance of being due
to an intelligent modification of the ancestral instinct ; but
how far the modification has become through heredity a

congenital variation, we do not know. The intelligence which is said to have enabled the martin of the past to adopt this method of nidification is still operative. The nestlings brought up under the eaves would have opportunities for acquiring experience which might lead them to build under similar circumstances. Nest and eaves would be associated in the conscious situation. Nor would the effects of natural selection be necessarily excluded. One may suppose that in the open country, far from rock-shelters, those martins in which there was a congenital tendency to build in house-shelters would bring up their broods and transmit this tendency; while those in which it was absent would either go elsewhere or fail to bring up broods at all. In the absence of fuller knowledge as to the truly instinctive nature of the behaviour, and as to its mode of genesis, we are in large degree at the mercy of conjecture. But in any case the incidence of elimination is not necessarily excluded, and there are, therefore, no grounds for denying that natural selection has been a co-operating factor in the evolution of the instinctive behaviour, if such it be.

It is well known that the lapwing will apparently simulate the actions of a wounded bird, with the object, as it seems, of drawing intruders away from her nest. And such tactics are not restricted to this bird, nor even to one or two species. They are common, no doubt with diversities of detail, to such different birds as grouse, pigeons, plovers, rails, avocets, ducks, pipits, buntings, and warblers. Granting that the behaviour is truly instinctive, it forms a very pretty subject for trans-missionists and their critics to quarrel over. "If we seek, as an·example," the transmissionist may exclaim, "an instinct which bears the marks of its intelligent, and therefore acquired origin, this of feigning wounded provides all that we can possibly demand." "What mode of instinctive behaviour," the selectionist may ask, "can be adduced which is more obviously useful to the species? Is not this just the kind of procedure which natural selection, if it be a factor at all, must fix upon and perpetuate through the elimination of failures? Those birds which, through congenital variation of behaviour, acted in this way would certainly enable their offspring to escape destruction by enemies, and these would survive to perpetuate the instinct."

Let us expand the transmissionist position a little further. An extremist, of the type presented by Eimer, would perhaps urge that the lapwing reasons thus : "If I pretend to be wounded, trail my wing, and flutter along the ground, instead of flying off, I shall draw upon myself the intruder's attention, and lead him to suppose that I shall be easily caught ; and if I thus entice him away, my little ones will be saved, and my end gained." Thus, it may be said, might the bird argue, and then give practical effect to its reasoning. But are we not here attributing to the lapwing powers of ratiocination beyond the capacity of the most intelligent of birds ? Are we not assuming a histrionic power, and a realization of the effects on others of its display, which many a human actor might well covet ?

"But may not the bird," it may be urged in reply, "have found by experience, without any elaborate process of abstract reasoning, that the trick is effectual ? " In any case it would be experience perilously acquired. Granting that the bird has the wit to try the trick, a little over-acting, a little too much lameness of wing, and she is herself seized and killed ; a little under-acting, and the trick fails—her brood is found and destroyed. Does it not seem probable that such experience would be dearly bought, that failure would mean either death to the parent or death to the offspring ? And is it not clear that natural selection is thus introduced in any case ? And may not the selectionist pertinently ask : "Why, if natural selection is thus introduced as a factor, halt midway between two hypotheses? Why not take the further step—one by which all the difficulties attending the intelligent acquisition and the biological transmission are alike avoided—of allowing that natural selection exercises, throughout, its influence on congenital variations, and not on acquired modifications of behaviour ? "

There is, however, a way in which, when natural selection is operative, intelligence may serve to foster congenital variations of the required nature and direction. We must remember that acquired habits on the one hand, and congenital variations of instinctive behaviour on the other hand, are both working, in their different spheres, towards the same end, that of adjustment to the conditions of life. If, then, acquired accommodation and congenital adaptation reach this end by different methods, survival may be best secured by their co-operation. And the more thorough-going the co-operation the better the chance of survival. There would be a distinct advantage in the struggle for existence when inherited tendencies of independent origin coincided in direction with acquired modifications of behaviour; a distinct disadvantage when such inherited tendencies were of such a character as to thwart or divert the action of intelligence. Thus any hereditary variations which coincide in direction with modifications of behaviour due to acquired habit would be favoured and fostered; while such variations as occurred on other and divergent lines would tend to be weeded out. Professor Mark Baldwin,* who has independently suggested such relation between modification and variation, has applied to the process the term "Organic Selection;" but it may also be described as the natural selection of coincident variations.

It may be urged, therefore, that if natural selection be accepted as a potent factor in organic evolution, and unless good cases can be adduced in which natural selection can play no part and yet habit has become instinctive, we may adopt some such view as the foregoing. While still believing that there is some connection between habit and instinct, we may regard the connection as indirect and permissive rather than direct and transmissive. We may look upon some habits as the acquired modifications which foster those variations which are coincident in direction, and which go to the making of instinct.

The net result of a study of instinctive behaviour is to lead

* Professor Henry Osborn has also indicated the relationship referred to.

us to the conclusion that its evolution runs parallel with the evolution of animal structure. This is perhaps best seen in the case of those insects in which typical instinctive acts are performed by larvæ of wholly different form and structure, though they are stages in the development of the same species. This is exemplified in the cases of *Sitaris*, *Argyromœba*, and *Leucopsis* which have been briefly described. It is probable that in all cases of instinctive action natural selection has been a co-operating factor. Without going so far as to assert with Professor Weismann the "all-sufficiency of natural selection," we may echo the words of Professor Groos,* and say : "Nevertheless, we know no principle except that of selection, and we must go as far as that will take us. Absolute knowledge of such phenomena is unattainable." And in this conclusion we have the support of Dr. Peckham, who says,† "We have found them [the instinctive acts of solitary wasps] in all stages of their development, and are convinced that they have passed through many degrees from the simple to the complex, by the action of natural selection. Indeed, we find in them beautiful examples of the survival of the fittest."

* "The Play of Animals," p. 64.
† "Solitary Wasps," p. 236.

CHAPTER IV

INTELLIGENT BEHAVIOUR

1.—The Nature of Intelligent Behaviour

Such an animal as a newly hatched bird or an insect just set free from the chrysalis is a going concern, a living creature. It is the bearer of wonderfully complex automatic machinery, capable, under the initiating influence of stimuli, of performing instinctive acts. But if this were all we should have no more than a cunningly wrought and self-developing automatic machine. What the creature does instinctively at first it would do always, perhaps a little more smoothly as the organic mechanism settled down to its work—just as a steam-engine goes more smoothly when it has been running for a while ; but otherwise the action would continue unchanged. Instinctive behaviour would remain unmodified throughout life. The chick, however, or the imago insect is something more than this. It affords evidence of the accommodation of behaviour to varying circumstances. It so acts as to lead us to infer that there are centres of intelligent control through the action of which the automatic behaviour can be modified in accordance with the results of experience. When, for example, a young chick walks towards and pecks at a ladybird, the like of which he has never before seen, the behaviour may be purely instinctive ; and so, too, when he similarly seizes a wasp-larva. Even when he rejects the ladybird or swallows the larva, this may be directly due to unpleasant stimulation in the one case, and pleasant in the other. But when, after a few trials, the chick leaves ladybirds unmolested while he seizes wasp-larvæ with increased energy, he affords evidence of selection based on individual experience. And such selection implies intelligence in almost its simplest expression. We may say, therefore, that, whereas instinctive behaviour is prior to individual experience, intelligent behaviour is the outcome and product of such experience. This distinction is presumably clear enough ; and it is one that is based on the facts of observation. But we must not fail to notice that, though the logical distinction is quite clear, the acquired modifications of behaviour, which we speak of as intelligent, presuppose congenital modes of response which are guided to finer issues. We may say, then, that where these congenital modes of response take the form of instinctive behaviour, there is supplied a general plan of action which intelligence particularizes in such a manner as to produce accommodation to the conditions of existence.

We have already frankly admitted that, in the present state of our knowledge, we do not know with any definiteness how intelligent modification of behaviour is effected. But it seems probable that from all parts of the automatically working organic machine messages come in to the centres of conscious control, and that in accordance with the net result of all these messages, past and present, tinged with pleasure or pain, other messages go out to the automatic centres, and, by checking their action here and enforcing it there, give new direction to the resulting behaviour. If this be so, the consciousness associated with the control-centres is like the person who sits in a central office and guides the working of some organized system in accordance with the information he is constantly receiving ; who sends messages to check activity in certain directions and to render it more efficient and vigorous in others.

It may be said, however, that intelligent guidance is, at any rate in such simple cases as the selection of a palatable grub and the rejection of a nauseous ladybird, itself determined by instinctive likes and dislikes. All young chicks apparently find wasp-larvæ palatable and ladybirds the reverse ; and this is just as much the outcome of heredity as the instinctive act of pecking. Since, therefore, heredity determines what shall be selected and what rejected—since the likes and dislikes are themselves instinctive—any essential difference between congenital and acquired behaviour seems to be evanescent.

Now, if we apply to the affective qualities of mental states —the pleasurable tone or its opposite which characterize such states—the term "instinctive," we do so in reference to the broader psychological conception of instinct, rather than in accordance with the narrower biological acceptation of the term. For the likes and dislikes constitute part of the conditions under which the behaviour occurs, and not elements in the co-ordinated response as such. Hence it is preferable to apply to these hereditary qualities the term *innate*, rather than the term *instinctive*. But, waiving this distinction, it is true that such pleasant or unpleasant qualities of the sensory results of stimuli are part of the animal's hereditary outfit, and are not acquired in the course of individual experience. What, then, is acquired ? What part does experience play in the development of intelligent behaviour ? Let us consider the case of the chick and the ladybird, and see whether it helps us to answer these questions. The chick is stimulated to the instinctive pecking response by a small moving object. That is the first scene of the little drama. In the second scene the ladybird is seized, sensory centres are unpleasantly stimulated, and the insect is dropped or thrown on one side with signs of disgust. Let us grant that this aversion with its characteristic response is also instinctive. *There is no hereditary connection between scene* 1 *and scene* 2. After an interval the curtain rises on act ii. The characters are the same as in the first scene of the previous act. But the action of the drama is different. The chick does not seize the ladybird. Why ? Because *there is an acquired connection between scenes* 1 *and* 2 *of the previous act.* The chick has gained experience of the nauseous character of the insect, and this experience influences and modifies his behaviour. The essentially new feature, therefore, is the establishment of a connection which is not provided through inheritance. To put the distinction in a brief form, we may say that instinct depends on how the nervous system is built through heredity; while intelligence depends upon how the nervous system is developed through use.

Assuming that an animal is capable of gaining experience and of acquiring new nervous connections in the course of individual experience, it follows that, as has already been indicated, instinctive behaviour in its logical purity is only presented in the first performance of any given co-ordinated act. For after this the animal has gained experience of its performance; and this can no longer conform to a definition of instinct, according to which it is characterized as "prior to experience." On the other hand, intelligent behaviour cannot be presented on the first occurrence of any action, since there is no prior experience thereof in the light of which it may be guided. This logical distinction may be expressed by saying that instinctive behaviour is always prior to experience, while intelligent behaviour is always subsequent to experience. When, however, instinctive procedure continues throughout life practically unmodified or but little modified, we may still class it under instinct, since the hereditary connections are still the predominant factors. And where the latter part of an instinctive sequence is modified by the experience gained in the former part, we may still term the modification intelligent, however small may be the time-interval implied in the word "subsequent." Sharp as the logical distinction is, the behaviour of animals is in the main a joint-product, and whether we term it instinctive or intelligent depends upon whether the hereditary or the acquired factor predominates.

Passing on now to consider some further characteristics of intelligent behaviour, we may first notice what Dr. Charles Mercier, in his work on "The Nervous System and the Mind," terms the four criteria of intelligence. Intelligence is manifested, he says, first in the novelty of the adjustments to external circumstances; secondly, in the complexity; thirdly, in the precision ; and fourthly, in dealing with the circumstances in such a way as to extract from them the maximum of benefit.

If, however, we are to regard these severally as criteria of intelligence, each should serve to differentiate intelligent from instinctive behaviour.   But this is not the case.   The precision of the adjustment cannot be regarded as a criterion of intelligence, for many instinctive acts are remarkably precise.   No grocer's assistant rolls a paper funnel with more precision than is displayed by the birch-weevil (*Rhynchites betulæ*) in constructing the leaf-case in which her eggs are laid.   Curved incisions of constant form are made on either side of the midrib, and are "of just the right shape to make the overlaps in the rolling, and to retain them rolled up with the least tendency

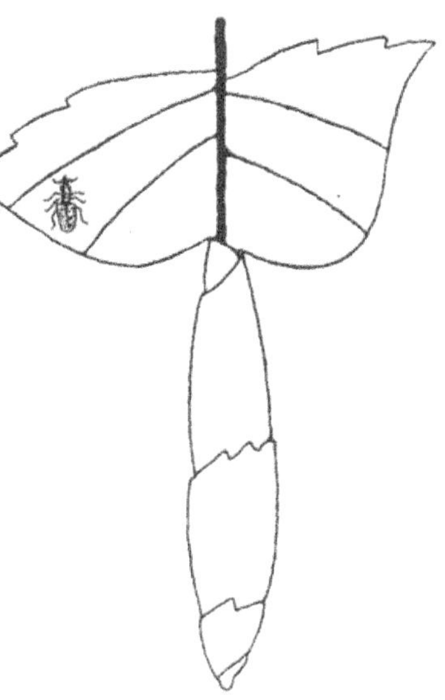

Fig. 18.—Leaf-case of Birch-weevil.

with the least tendency to spring back," * while the tip of the leaf is rolled into a second smaller funnel, which is tucked in to close the opening of the first, after the eggs have been deposited.   "The eggs hatch in their dark place, each giving rise to an eyeless maggot, which ultimately leaves the funnel for the earth. . . . Hence the beetle cannot be considered to have ever seen a funnel, and certainly has never witnessed the construction of

* See D. Sharp, "Cambridge Natural History," "Insects," part ii., p. 293, and the original authorities cited on p. 294.

one, though, when disclosed, it almost immediately sets to
work to make funnels on the complex and perfect system "
characteristic of the species. This is but one example of
instinctive precision out of the many which could be cited.
We may say, then, that though, when an act is otherwise
shown to be intelligent, the precision is a criterion of the
level attained by the intelligence, still it cannot be said to
be a criterion which serves to distinguish intelligent from
instinctive behaviour.

Nor can we regard apparent prevision (which is sometimes
advanced as a criterion of intelligence) as specially distinctive
of intelligent acts regarded objectively in the study of animal
behaviour. For, as we have had occasion to show, there are
many instincts which display an astonishing amount of what
may be termed "blind prevision"—instance the instinctive
regard for the welfare of unborn offspring which the mother
will never see, and the instinctive preparation for an unknown
future existence in the case of insect larvæ.

Nor, again, is the complexity of the adjustment distinctive
of intelligence as contrasted with instinct. We have cited
examples which afford evidence of much complexity in instinctive behaviour. The construction and storage of the
nest among solitary wasps, and their methods of capturing
and conveying the insects or spiders on which they prey, are
sufficiently complex. So, too, is the behaviour of the Sitaris
larva which attaches itself to the male bee, passes to the
female, and then slips on to the eggs she lays ; and so, again,
is that of the Yucca moth, which collects pollen from the
anthers, conveys it to the stigma, and then lays her eggs
among the ovules. These cases show, too, that the circumstances may be dealt with in such a way as to extract from
them the maximum of benefit. It would be difficult intelligently to improve upon the manner of dealing with the circumstances displayed in many familiar modes of instinctive procedure.

There remain the novelty of the adjustment and the individuality displayed. And here we seem to have valid
criteria of intelligent behaviour. The ability to perform

acts in special adaptation to new circumstances, and the individuality manifested in dealing with the complex conditions of a variable environment,—these seem to be distinctive features of intelligence. On the other hand, in instinctive behaviour there seems to be no choice ; the animal is impelled to their stereotyped performance through impulse, as by a stern necessity ; they are so far from novel that they are performed by every like individual of the species, and have been so performed by their ancestors for generations ; and in performing the instinctive act, the animal seems to have no more individuality or originality than a piece of adequately wound clockwork.

Granting, then, that behaviour is shown to be intelligent by the fact that there is evidence of profiting by experience, we may say that the level attained by the intelligence is indicated by the complexity of the adjustment, its precision, the individuality shown, the amount of prevision disclosed, and in its being such as to extract from the circumstances the maximum of benefit. Many of these points, however, serve equally well to mark the level of instinctive procedure.

II.—Intelligent Behaviour in Insects

It is, as we have seen, among the higher invertebrates, especially in insects, that some of the most remarkable and complex instincts may be found. There is,* however, a tendency to ascribe the behaviour of insects entirely to instinct, without sufficient evidence that neither imitation, instruction, nor intelligent learning play any part. This is, perhaps, a survival of the old-fashioned view that all the acts of the lower animals are performed from 'instinct, whereas those of human beings are to be regarded as rational or intelligent. In popular writings and lectures, for example, we frequently find some or all of the following activities of ant-life ascribed to instinct :

* This and the three succeeding paragraphs are taken from " Animal Life and Intelligence," p. 425.

recognition of members of the same nest; powers of communication; keeping aphides for the sake of their sweet secretion; collection of aphid eggs in October, hatching them out in the nest, and taking them in the spring to the daisies, on which they feed, for pasture; slave-making and slave-keeping, which, in some cases, is so ancient a habit that the enslavers are unable even to feed themselves; keeping insects as beasts of burden, *e.g.* a kind of plant-bug to carry leaves; keeping beetles, etc., as domestic pets; habits of personal cleanliness, one ant giving another a brush-up, and being brushed-up in return; habits of play and recreation; habits of burying the dead; the storage of grain and nipping the budding rootlet to prevent further germination; the habits described by Dr. Lincecum, and to a large extent confirmed by Dr. McCook, that Texan ants prepare a clearing around their nest, and six months later harvest the ant-rice, a kind of grass of which they are particularly fond, even, according to Lincecum, seeking and sowing the grain which shall yield this harvest; the collection by other ants of grass to manure the soil on which there subsequently grows a species of fungus upon which they feed; the military organization of the ecitons of Central America; and so forth. Now, the description of the habits of ants forms one of the most interesting chapters in natural history. But to class them all as illustrations of instinct is a survival of an old-fashioned method of treatment.

To put the matter in another way. Suppose that an intelligent ant were to make observations on human behaviour as displayed in one of our great cities or in an agricultural district. Seeing so great an amount of routine work going on around him, might he not be in danger of regarding all this as evidence of hereditary instinct? Might he not find it difficult to obtain satisfactory evidence of the establishment of our habits, of the fact that this routine work has to some extent to be learnt? Might he not say (perhaps not wholly without truth), "I can see nothing whatever in the training of the children of these men to fit them for their life-work. The training of their children has no more apparent bearing upon the activities of their after-life than the feeding of our grubs has on the duties of ant-life. And although we must remember," he might continue, " that these large animals do not have the advantage which we possess of awaking suddenly, as by a new birth, to their full faculties, still, as they grow older, now one and now another of their deferred instincts is unfolded and manifested. They fall into the routine of life with little or no training as the period proper to the various instincts arrives. If learning thereof there be, it has at present escaped our observation. And such intelligence as their activities evince (and many of them do show remarkable adaptation to uniform conditions of life) would seem to be rather ancestral than of the present time ; as is shown by the fact that many of the adaptations are directed rather to past conditions of life than to those which now hold good. In the presence of new emergencies to which their instincts have not fitted them, these poor men are often completely at a loss. We cannot but conclude, therefore, that, although shown under somewhat different and less favourable conditions, instinct occupies fully as large a space in the psychology of man as it does in that of the ant, while their intelligence is far less unerring and, therefore, markedly inferior to our own."

Of course, the views here attributed to the ant are very absurd. But are they much more absurd than the views of those who, on the evidence which we at present possess, attribute all the varied activities of ant-life to instinct ? Take the case of the ecitons, or military ants, or the harvesting ants, or the ants that are said to keep draught-bugs as beasts of burden : have we sufficient evidence to enable us to affirm that these modes of behaviour are purely instinctive and not intelligent ; that all the varied manœuvres of the military ants, for example, are displayed to the full without any learning or imitation, without teaching and without intelligence on the part of every individual in the army.

That in some cases there is something very like a training or education of the ant when it emerges from the pupa condition is rendered probable by the observations of M. Forel.

As Romanes says,* "The young ant does not appear to come into the world with a full instinctive knowledge of all its duties as a member of a social community. It is led about the nest and 'trained to a knowledge of domestic duties, especially in the case of larvæ.' Later on, the young ants are taught to distinguish between friends and foes."

We have only to weigh the evidence brought forward by such observers as Fabre and Dr. Peckham to see that among the solitary wasps and mason bees the behaviour, though founded on instinct, is in large degree modified by intelligence. The care with which a site for the tunnelled nest in the ground is selected, betokens something more than instinct. The following is a slightly condensed statement of Dr. and Mrs. Peckham's observations on one of the solitary wasps (*Aporus fasciatus*).† "We were working one day in the melon-field when we saw one of these little wasps going backwards and dragging a spider. She twice left it on the ground while she circled about for a moment, but soon carried it up on to one of the large melon leaves, and left it there while she made a long and careful study of the locality, skimming close to the ground in and out among the vines; at length she went under a leaf close to the ground, and began to dig. After her head was well down in the ground we broke off the leaf that we might see her method of work. She went on for ten minutes without noticing the change, and then, without any circling, flew off to visit her spider. When she tried to return to her hole it was evident that some landmark was missing. Again and again she zig-zagged from the spider to the nesting-place, going by a sort of path among the vines from leaf to leaf, and from blossom to blossom, but when she reached the spot she did not recognize it. At last we laid the leaf back in its place over the opening, when she at once went in and resumed her work, keeping at it steadily for ten minutes longer. At this point she suddenly reversed her operations, and began to fill in the hole that she had made. She then glanced at the spider,

* "Animal Intelligence," p. 59.
† "Instincts and Habits of the Solitary Wasps," p. 55.

selected a new place, and began to dig again. This hole was also filled in ; she looked once more at the spider, and started a nest in a new place. This, in turn, was soon abandoned, as was a fourth. The fifth beginning was made under a leaf that lay close to the ground, but after twenty minutes' work this place also was abandoned and a sixth started. This, however, was the final choice, and after forty-five minutes spent in digging it was completed."

This description shows an amount of apparent fastidiousness which is quite irreconcilable with the hypothesis that the behaviour is merely instinctive. Not less fastidious are some wasps in the temporary closure of the hole with a stone or pellet of earth, the operation being repeated several times

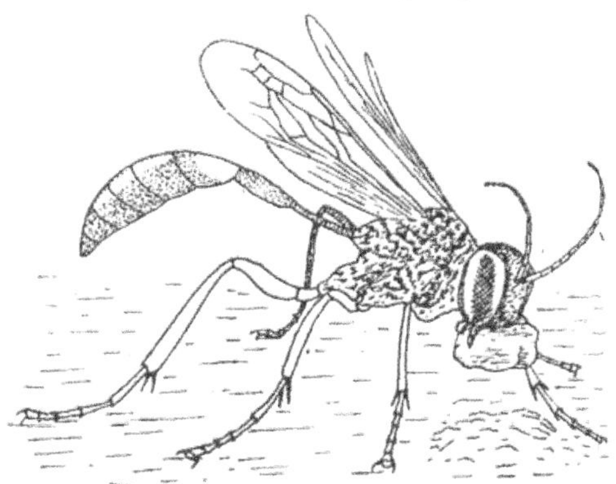

Fig. 19.—Solitary Wasp using a stone to beat down the earth over its nest (after Peckham).

with different covers before the insect seems to be satisfied ; while in other cases the hole is hidden by bringing earth in such quantity as to render the place indistinguishable from the rest of the field. But in one case observed by Dr. Peckham, intelligent procedure was carried so far as apparently to involve the use of a tool, the same behaviour having been independently observed in the same genus (*Ammophila*) by Dr. S. W. Williston of Kansas University. "Just here," writes Dr. Peckham,* "must be told the story of one little wasp whose individuality stands out in our minds more distinctly than that of any of the others. In filling up her nest she put her head down into it and bit away the loose earth from the sides, letting it fall to the bottom of the burrow, and then, after a quantity had accumulated, jammed it down with her head. Earth was then brought from the outside and pressed in, and then more was bitten from the sides. When, at last, the filling was level with the ground, she brought a quantity of fine grains of dirt to the spot, and, picking up a small pebble in her mandibles, used it as a hammer in pounding them down with rapid strokes, thus making this spot as hard and firm as the surrounding surface. Before we could recover from our astonishment at this performance she had dropped her stone and was bringing more earth, and in a moment we saw her pick up the pebble and again pound the earth into place with it. Once more the whole process was repeated, and then the little creature flew away."

Here we have intelligent behaviour rising to a level to which some would apply the term rational. For the act may be held to afford evidence of the perception of the relation of the means employed to an end to be attained, and some general conception of purpose. In this section, which deals with description of behaviour based on observation, the psychological explanation cannot be discussed. Similar indications of deliberate action may be held to be afforded by the sometimes elaborate "locality studies" which these insects seem to make,—by the "care that is taken by wasps to acquaint themselves with the surroundings of their nests." A *Sphex*, for example, which had partially made and then abandoned several nests, left them without any locality study; but when she had completed a nest in a suitable spot she made "a most thorough and systematic study of the surroundings. She flew in and out among the plants, first in narrow circles near the surface of the ground, and then in wider and wider ones as

* *Op. cit.*, p. 22.

she rose higher in the air, until at last she took a straight line and disappeared in the distance." Another species (*Cerceris deserta*) "has the habit of making a number of half circles in front of the nest, and then, after rising a little higher, of flying several times completely round it." The method of procedure is, it seems, so normal to the species that it is probably founded on an instinctive basis. Dr. and Mrs. Peckham, in commenting on their observations, say : * "If the examination of the objects about the nest makes no impression upon the wasp, or if it is not remembered, she ought not to be inconvenienced nor thrown off her track when weeds and stones are removed and the surface of the ground is smoothed over ; but this is just what happens." For convenience of observation they "sometimes gently moved intercepting objects to one side, but even such a slight change threw the wasp out of her bearings, and made it difficult for her to recover her treasure." Where wasps form a number of nests in a small plot of ground, as in the case of *Bembex*, each knows and returns to its own hole, as was proved by Dr. Peckham, who marked the insects and their nests with paint.

So, too, with regard to prey. In the course of his observations on *Pompilus*, Fabre removed the spider which the wasp had deposited on a tuft of vegetation before she made her nest. As she was at work beneath the surface she could not see what went on above ground or where the spider had been redeposited some twenty inches from its former position. On emerging from the nest the wasp went straight to the original spot, searched round it for some time, then made further excursions, and discovered the spider. After slightly altering its position, and placing it on another tuft of vegetation, she returned to her subterranean labours, giving the observer an opportunity of again moving the spider. Five times did Fabre repeat the operation, and every time the wasp returned to the spot where she had last deposited her prey.

The same observer records some interesting experiments with the mason bee, *Chalicodoma*. The mud nests of the species

* *Op. cit.*, p. 215.

K

investigated were affixed to stones on the banks of the Rhone. When a nest was partially constructed, the bee having flown off for more material, Fabre moved the stone to a new position, near at hand and easily visible from the original site. The bee went straight to the place where the nest had been, searched the immediate neighbourhood, flew off, and returned to the same spot to continue the search. If she came across her own nest in its new position she did not recognize it as hers, but left it after examination. But if a stone with the nest of another bee in about the same stage of construction was placed in the position occupied by her own, she adopted it. And when two nests near together, both half built, were transposed, each bee unhesitatingly adopted the nest which occupied the position where its own nest had been. It may well seem strange that, the general locality-memory being so well marked, the recognition of the particular stone and nest should be deficient. This may be due to the fact that the so-called compound eyes are the organs concerned in locality vision, while the ocelli deal with details at very close range, and that the former alone afford the requisite data for recognition; by their instrumentality alone arises the conscious situation which affords guidance in behaviour. And in that situation slight changes which for us make it "still the same but with a difference" render it no longer the same for a being of more limited intelligence—one probably incapable of analyzing the situation and seeing that the sameness preponderates over the difference. Be this as it may, the failure of a bee to recognize its own nest under circumstances so foreign to its experience as removal to a new spot may be paralleled with what I have observed in the case of sticklebacks. A nest had been built in a round glass bell jar which stood near a window. Some aquatic vegetation grew in the tank, and the nest was built on the window side. An experiment was made by turning the large bell jar through a right angle. The male stickleback searched for its nest in the old direction on the window side—that is to say, the same position in reference to the incidence of the light. The search was, of course, fruitless, and a new nest was begun in this position. Presently the old nest was discovered, and was then vigorously destroyed in just the same way as the nest of a rival is pulled to pieces and scattered. Here a new incidence of light and new direction of shadows seemed to have completely transformed the visual situation.

To return to insects, it is probable that the homing faculty is not the result of an inborn mysterious instinct dependent on some sense of direction of which we have no knowledge, but is based upon experience gained during their flight hither and thither—that, in a word, it is intelligent and not instinctive. Experiments of Fabre at first seemed to suggest some magnetic influence to which bees were sensitive ; for when a minute magnet was fixed to a bee as it started on its return journey, the insect was at fault ; but as a check experiment he affixed a piece of straw instead of a magnet, with similar results. Some of Fabre's observations and those of Dr. Bethe * are difficult to reconcile with the hypothesis that, in the homing, guidance is due to acquired acquaintance with the locality. But, on the other hand, the experiments of Lord Avebury and of Romanes seem to favour this view. Romanes found that when bees were taken inland from their hive near the seaboard, and then liberated, they returned from considerable distances, the whole locality being familiar ; but taken to the seashore, where the objects around them were unfamiliar (since the seashore is not the place where flowers and nectar are to be found), the bees, though not far distant from the hive, were nonplussed and lost their way. Dr. and Mrs. Peckham, as the result of their extremely careful observations, seem to have no doubt that the homing of solitary wasps is due to locality-experience ; and of the social wasp, *Polistes fusca*, they say : † " We have seen the young workers make repeated locality studies when they first began to venture

* " Dürfen wir den Ameisen und Bienen psychische Qualitäten Zuschreiben." *Pfluger's Archiv.*, lxx., 1898.

† " Solitary Wasps," p. 219.

away from home, but as they occupy the same nest all summer they, of course, grow more and more familiar with their surroundings, until they become so thoroughly acquainted with them that they can find their way without the least difficulty. We have no doubt that with them, as with the solitary wasps, the faculty is not instinctive, but is the direct outcome of individual experience."

In the interesting pages of the works in which Dr. and Mrs. Peckham describe their investigations, there are many observations which show that wasps are capable of intelligently profiting by the experience which their instinctive behaviour places them in a position to acquire. The inherited tendencies and aptitudes pave the way for acquired modification and accommodation of behaviour. To catch and paralyze spiders, to dig and prepare a tunnelled nest, and to carry the prey to the nest, all this affords the instinctive basis ; but when the observers tell us that they " have several times seen wasps enlarge their holes when a trial had demonstrated that a spider would not go in," and even on one occasion without trial when an unusually bulky spider was brought, there is something beyond instinct ; there is intelligent adjustment to special circumstances given in experience. Presumably intelligent is the habit frequently observed in one species of *Pompilus*, and occasionally in another, of hanging the paralyzed spider in a crotch of a branching stem, usually of bean or sorrel, where it will be safe from the depredation of ants. On one occasion Dr. Peckham, desirous of seeing the exact manner in which the victim was stung, substituted an unhurt spider for that which the wasp had paralyzed.[*] " According to the habit of its species when danger threatens, it kept perfectly quiet, and when the wasp returned it was hanging there as motionless as a piece of dead matter ; but she would not touch it ; she hunted all over that plant and then over several others near to it, returning continually to look again at the right spot. After five minutes she flew off in the direction of the woods to catch another spider. Why did she go to the woods ? Why did she not

* *Op. cit.*, pp. 131, 132.

take the one that hung there in plain view ? It could not
have been due to the fact that we had handled the spider,
since when, on other occasions, we took one that had been
paralyzed, examined it, and then returned it to the wasp, she
accepted it without hesitation. . . . In forty minutes she came
back with another spider, but, instead of taking it into the nest,

Fig. 20.—Spiders placed by Solitary Wasps in the crotches of
branching stems (after Peckham).

she hung it upon a bean plant near by, and then proceeded to
dig a new hole a few inches distant from the first. Foolish
little wasp, what a waste of labour ! Truly, if you are endowed
with energy beyond your fellows, you are but meagrely furnished
with reason."

Here we have the routine of instinct—the normal mode of
hunting and capturing prey, the normal procedure of bringing
the spider, and then making the nest, predominating over any
tendency to initiate intelligent improvements. This, however,
should not surprise us, in whom the force of habit is often so
strong. Nor should we feel surprise at the apparently stupid
tolerance some of these wasps display in presence of parasites.

Bembex, which does not store and close its cell, but brings
continual supplies of food to its larvæ, is not disturbed by the
presence in the cell of the grubs of the parasitic fly *Miltogramma*. She could, we think, easily free her nest of these
intruders, but she continues to bring supplies, though the
parasites may absorb it all and leave her own larvæ to perish.
She adapts her procedure to the new conditions, being incapable of knowing that she is feeding the enemies of her
race.

Enough has now been said to show the extent and the
limitations of the intelligence of such insects as the solitary
wasps.   It will be noticed that the acquired modifications of
behaviour occur in close connection with the inherited ground-
plan of instinctive procedure.   We shall have occasion to note
the same connection in our discussion of social behaviour in
the next chapter.   And we shall consider the influence of
intelligence on instinct before we bring this chapter to a
close.

III.—Some Results of Experiment

It is unnecessary to give a *resumé* of entertaining anecdotes
illustrative of intelligent behaviour in the higher animals.
Such anecdotes are too often the outcome of casual observations by untrained observers ; and the interpretation put upon
the facts frequently shows a want of psychological discrimination.   Such is not the material of which science is constituted.
What is needed is systematic observation conducted, so far as
possible, under controlled conditions.   Two things are necessary : first, to distinguish instinctive behaviour, inherited as
such, from the acquired modifications or new departures due
to intelligence ; and secondly, to determine the method and
range of intelligent procedure.   These problems can only be
solved in their entirety by a complete knowledge of the life-
history of the animal concerned.   But they may be attacked
in detail by a systematic study of particular modes of behaviour,
and by an investigation into their manner of origin.   That

this may be done with some approach to accuracy, resort must be had to experiment, which permits of observation under controlled conditions.

To ascertain, for example, how far nest-building is instinctive in birds, Mr. John S. Budgett hatched a hen greenfinch under a canary. In the following autumn he bought a caged bird, a cock, probably of the same year, and in the succeeding spring turned the pair into a large aviary, supplying such material as twigs, rootlets, dried grass, moss, feathers, sheep's wool and horsehair. The hen soon began to build, the cock bird taking no share in the work, and finished her nest in a few days. On careful comparison it was found to resemble that of a wild greenfinch in every particular, being made of wool, roots, and moss, lined with horsehair. A second nest the aviary greenfinch built was also quite normal.

In the case of a bullfinch which Mr. Budgett reared, having obtained it when a few weeks' old, the first nest was composed of dried grass with a little wool and hair, but without either rootlets or twigs. A second which she built was, however, quite typical, made of fine twigs and roots, and lined with horsehair ; as was also a third nest.

It is just possible, though most improbable, that the bullfinch utilized its three weeks' experience gained in the nest from which it was taken. But Mr. Jenner Weir describes [*] observations on canaries in which this source of experience is excluded. " It is usual," he says, " with canary fanciers to take out the nest constructed by the parent birds, and to place a felt nest in its place, and when her young are hatched, and old enough to be handled, to place a second clean nest, also of felt, in the box, removing the other. But I never knew that canaries so reared failed to make a nest when the breeding time arrived. I have, on the other hand, marvelled to see how like a wild bird's the nests are constructed. It is customary to supply them with a small set of materials, such as moss and hair. They use the moss for the foundation, and

* In a letter to Darwin, quoted by Romanes, "Mental Evolution in Animals," p. 226.

line with the finer materials, just as a wild goldfinch would
do."

Experiment seems, therefore, to show in a way, and with a
clearness impossible of attainment by observation under natural
uncontrolled conditions, that nest-building in birds is instinctive. That the manner and method of procedure is often
modified in accordance with special conditions—that the instinctive outline of nidification receives its final touches
through individual experience—is sometimes seen under
nature, and more often under the semi-experimental conditions
of domestication. Thus three pairs of pigeons in the Wilson
Tower of Clifton College made their nests in 1898, as I am
informed by Mr. H. C. Playne, of galvanized iron wire, pieces
of which were left in a corner at the top of the tower, thus
affording a parallel to the behaviour of the unconventional
crow of Calcutta, mentioned by Mr. J. W. Headley,* which
made its nest of soda-water bottle wires, which it picked up in
a back yard. But even in this matter experiment serves to
bring out clearly the selective influence which is exercised by
intelligence. Bolton,† in 1792, observed a pair of goldfinches
beginning to build their nest in his garden. They formed the
ground-work of moss, grass, etc., as usual ; but on his scatter-
ing small parcels of wool in different parts of the garden they,
in great measure, left off the use of their own stuff and used
the wool. Afterwards he gave them cotton, and they then
used this instead of the wool ; then he supplied fine down, and
they finished their work with this, leaving the wool and
cotton.

In studying the behaviour of wild animals under natural
conditions, it must always be difficult to distinguish the congenital basis from the acquired elements ; for both tend to
bring about a working adjustment to the conditions of life,
and we can seldom have opportunities of tracing the interplay
of the factors which produce the instinct-habits of adult life.

---

* "The Structure and Life of Birds," p. 335.

† Preface to "Harmonia Ruralis," quoted by Yarrell, "British Birds,"
vol. i., p. 541.

But under domestication we seek to bring about a new working adjustment to conditions imposed by man. The skilful trainer utilizes the natural instinctive tendencies as a basis ; and, by a system of rewards and punishments, leads the intelligent modifications of behaviour along lines directed by his deliberate purpose. The conditions are largely those of experiment, and they bring out the play and range of intelligence in a way which would otherwise elude our observation. The training of falcons for the chase affords a good illustration, since they cannot be bred in confinement, and the effects of training cannot therefore be hereditary. The falconer's object is to modify the congenital instinctive behaviour of a bird of prey for the purposes of sport. She is trained to the lure at first at short distances, and step by step through longer flights ; she is taught by snatching away the lure to stoop at it repeatedly as often as it is jerked aside ; and then she is trained on living quarry, at first under easy conditions, till eventually she can be flown at a wild bird. And as a result a well-trained falcon will follow her master from field to field, regulating her flight by his movements, always ready for a stoop when the quarry is sprung. The fact that she can be thus educated for her work shows that her behaviour is plastic, and can be moulded by intelligence. Experimental conditions reveal the fact ; but under nature the moulding influence of intelligence is presumably not less important, though it is more directly in line with the congenital instinctive tendencies.

That much of the behaviour of the higher animals is guided by experience similar to that which plays so large a part in their training under the experimental conditions of domestication is generally admitted. But what are the range and limits of animal intelligence, and whether it attains the level of rational conduct, in the restricted sense of the term " rational," are questions open to discussion, to which answers are more likely to be obtained through experiment than by chance observations.

Before giving some of the results of such experiment it will be well to revert to the distinction, which was drawn in the second chapter, between the lower or intelligent stage of mental development and the higher or rational stage. It will be remembered that rational processes were characterized by the fact that the situations contain the products of reflective thought, presumably absent in the earlier stages of development ; that they were further characterized by a new purpose or end of consciousness, namely, to explain the situations which at an earlier stage are merely accepted as they are given in presentation or re-presentation ; they require deliberate attention to the relationships which hold good among the several elements of successive situations ; and they involve, so far as behaviour is concerned, the intentional application of an ideal scheme with the object of rational guidance.

On the other hand, the animal at the stage of intelligent behaviour deals with the circumstances of his comparatively simple life by making use of the particular situations which have been presented to consciousness in the course of his practical experience. If such an animal be placed in the midst of new circumstances he has to find out by a process of trial and error how they are to be met. After a longer or shorter period of trial, guided only by particular experiences, he chances to hit upon a mode of procedure which is successful. The successful act is then incorporated in a new situation ; at first, perhaps, only incompletely. The association is eventually established by repetition, through which is acquired the habit of doing the right thing in the appropriate manner. Why he does this and not something else, in so far as he is intelligent and not rational, he probably neither knows nor has the wit to consider. The satisfaction of success suffices for intelligence as such. If the circumstances be so modified as to render the particular mode of meeting them ineffectual, after trying again and again in the old way, he will sometimes stumble upon the proper mode of overcoming the difficulty, and after doing so two or three times a new conscious situation involving the requisite associations will be established, and the appropriate behaviour will become habitual. But why this new mode of procedure rather than any other is adopted, intelligence as such does not know, because it does not analyze the situation and disentangle the essential relationships. The satisfaction of success again suffices. In a word, such an animal in the perceptual stage of mental development seems wanting in the power of reflection. He does not appear to show evidence of framing anything like a general scheme of knowledge which he can apply to the solution of particular problems, of a practical nature, involving difficulties and obstacles.

The method of intelligence—in the sense in which we are using this term—the method of varied trial and error with the utilization of chance success, is a lengthy and somewhat clumsy process; but it suffices. Now contrast it with the procedure of a rational animal, such as man is or may be. When he is confronted by a difficulty he is not content to meet it by trying this way, and that way, and another way, anyhow, and trusting to chance to bring success, but he considers the problem in all its relations with a view to ascertaining the essential nature of the difficulty. For each attempted mode of meeting the case he has a definite reason. He knows why he does this and not that. He has a plan or scheme which he puts into execution. And if it fail, he is not content till he finds out wherein the failure lay. This enables him to plan a better scheme. He sees why it is better; and if at last he be successful by a happy hit, as in the chance procedure of intelligence, he looks for the reason of it. And seeing why this fortunate attempt, unlike his previous efforts, just meets the case, he repeats it because he perceives that herein lies the essential solution of the difficulty. Both in the case of intelligence and in that of reason, as here distinguished, present procedure is based upon past experience; but reason has built upon the foundations thus laid an orderly scheme, and knows its whys and wherefores, while intelligence is at the mercy of chance associations. The reason for success it has not the wit to assign.

The essential difference between the two cases may be put in another way by saying that the intelligent being forms sensory impressions and sensory images linked together by bonds of association, combining and coalescing to constitute a conscious situation effective in behaviour under the guiding influence of pleasure and pain ; while the rational being not only does all this, but goes further. He fixes his attention on the way in which the elements in the situation are connected and related; he builds an ideal framework on which the sensory impressions are set or move in an orderly manner. And it is this scheme, fashioned by reason and transforming the situation, which he utilizes in dealing with difficulties.

Yet another way of putting the same essential distinction is to say that intelligence deals with pictures, either directly presented to the senses or called up in re-presentation. If we state the matter thus, however, we must remember that the " pictures " may be painted in colours supplied by any of the senses ; and that smells, tastes, sounds, touches, pressures, limb-movements, and so forth, are elements in the pictured product. Bearing this in mind, we may say that intelligence deals with sensory impressions and their revived images in concrete and particular situations ; while reason analyzes the pictures, and extracts from them general notions in terms of which the pictures may be explained. For example, we picture a stone falling to the earth ; but we explain it by the general notion of gravitative attraction. The conception forms part of our ideal scheme of knowledge, which is not itself picturable, though this or that example of its action may be presented or re-presented in sensory imagery.

Once more we may say—and this way of looking at the question arises naturally out of what has gone before—that intelligence deals with concrete examples, and does not rise to the abstract and general rule. The ideal scheme of reason is the result of abstraction and generalization. It is a framework of conceptions which can be applied to the particular facts which fall under observation to see whether it fits and meets the case. Intelligence has to deal with the facts as they present themselves, without the aid of an organized system of knowledge built up into an ideal scheme.

Enough has now been said to indicate the distinction

between the method of intelligence and that of reason. It may, no doubt, be said that the terminology used is open to criticism ; for, on the one hand, the words "intelligence" and "intelligent" are frequently used as synonymous with "reason" and "rational ; " and, on the other hand, acts requiring neither abstraction, generalization, nor the application of any scheme of knowledge are frequently spoken of as "rational." Hence there is, it may be urged, some danger of misunderstanding. This may be granted. And unless some such restriction of meaning under suitable terms be accepted by psychologists, misunderstanding will continue. More essential, however, than the distinctive terms we are to use is the distinction of method which underlies them. That, I trust, is sound. Dr. Lindley, in an interesting paper on "A Study of Puzzles," * has utilized the distinction in his investigation of the mental development of children, and has found that the procedure of young children is predominantly of the "sense-trial and error" order which has above been termed intelligent ; and he expresses the opinion that "most of the adaptations of animals are on this sense-trial and error level."

Such certainly seems to be the conclusion to be drawn from my own experimental observations on dogs. It has frequently been asserted that the behaviour of a dog with a stick in his mouth, when he comes to a narrow gap, shows that he at once perceives the nature of the difficulty, and meets it in a rational manner by adopting the appropriate plan of action. He pulls the stick through by one end. But experiments, which I have elsewhere described,† showed that a fox terrier, fourteen months old, seemed to be incapable of perceiving the nature of the difficulty which vertical iron railings presented to his passage with a stick in his mouth, and only imperfectly learnt to overcome it after many ineffectual trials and many failures. The results obtained on the first afternoon may be quoted to indicate the nature of the evidence. The dog was sent after a short stick into a field, and had to pass through vertical rails about six

---

* *American Journal of Psychology*, vol. viii., no. 4, pp. 431–493.
† "Introduction to Comparative Psychology," p. 255.

inches apart. On his return the stick caught at the ends. I whistled and turned as if to leave; and the dog pushed and struggled vigorously. He then retired into the field, lay down, and began gnawing the stick, but, when called, came slowly up to the railings and stuck again. After some efforts he put his head on one side, and brought the stick, a short one, through. After patting and encouraging him, I sent him after it again. On his return he came up to the railings with more confidence, but, holding the stick by the middle, found his passage barred. After some struggles he dropped it and came through without it. Sent after it again, he put his head through the railings, seized the stick by the middle, and then pulled with all his might, dancing up and down in his endeavours to effect a passage. Turning his head in his efforts, he at last brought the stick through. A third time he was again foiled; again dropped the stick; and again seizing it by the middle tried to pull it through. I then placed the stick so that he could easily seize it by one end and draw it through the opening between the rails. But when I sent him after it, he went through into the field, picked up the stick by the middle, and tried to push his way between the railings, succeeding, after many abortive attempts, by holding his head on one side.

Subsequent trials on many occasions yielded similar results. But the following summer, when I resumed the experiments, I was able with some guidance to teach him to bring a long stick to the railings, drop it, and then draw the stick through by one end; though even then, if he had dropped it so that one end just caught a rail, he often failed, shaking his head vigorously, dropping the stick and seizing it again, and repeating this behaviour until it chanced to fall in a more favourable position. He did not apparently perceive that by gently moving the stick a little one way or other the difficulty could be simply overcome with little effort. Nor when given a crooked stick, which caught in a rail, did he show any sign of perceiving that by pushing the stick and freeing the crook he could pull the stick through. Each time the crook caught he pulled with all his strength, seizing the stick now at the end, now in the middle, and now

near the crook.   At length he seized the crook itself, and with
a wrench broke it off.   A man who was passing, and who had
paused for a couple of minutes to watch the proceedings, said,
" Clever dog that, sir ; he knows where the hitch do lie."   The
remark was the characteristic outcome of two minutes' chance
observation.   During the half hour or more during which I
had watched the dog he had tried nearly every possible way of
holding and tugging at the stick.   Such is the mode of
behaviour based on intelligence—continued trial and failure,
until a happy effect is reached, not by methodically planning,
but by chance.

Two of my friends criticized these results, and said that
they only showed how stupid *my* dog was.   *Their* dogs would
have acted very differently.   I suggested that the question
could easily be put to the test of experiment.   The behaviour
of the dog was in each case—the one a very intelligent York-
shire terrier, the other an English terrier—similar to that
above described.   The owner of the latter was somewhat
annoyed, used forcible language, and told the dog that he
could do it perfectly well if he tried.

In experimenting with my fox terrier on the method
adopted in seizing and carrying differently balanced objects,
I used (1) a straight stick, the centre of gravity of which was
at the middle ; (2) a Kaffir knob-kerrie, the centre of gravity
of which was about six inches from the knob ; (3) a light
geological hammer ; and (4) a heavier hammer.   In the last,
the centre of balance was close to the hammer head.   The net
result of the observations was that the best place for seizing
and holding the object was hit upon in each case after indefinite
trials ; that after three or four days' continuous experience with
one (say the knob-kerrie), another (say the stick) was at first
seized nearer one end, showing the influence of the more recent
association ; and that there was little indication of the dog's
seizing any one of the four at once in the right place, that is
to say, the point of seizure was not clearly differentiated in
accordance with the look of the object.   I tied a piece of
string, in later trials, round the centre of balance, but this, at

the time of the dog's death, had not served as a sure guide to his experience.

The way in which my dog learnt to lift the latch of the garden gate, and thus let himself out, affords a good example of intelligent behaviour. The iron gate outside my house is held to by a latch, but swings open by its own weight if the latch be lifted. Whenever he wanted to go out the fox terrier raised the latch with the back of his head, and thus released the gate, which swung open. Now the question in any such case is: How did he learn the trick? In this particular case the question can be answered, because he was carefully watched. When he was put outside the door, he naturally wanted to get out into the road, where there was much to tempt him—the chance of a run, other dogs to sniff at, possibly cats to be worried. He gazed eagerly out through the railings on the low parapet wall shown in the illustration; and in due time chanced to gaze out under the latch, lifting it with his head. He withdrew his head and looked out elsewhere; but the gate had swung open. Here was the fortunate occurrence arising out of natural tendencies in a dog. But the association between looking out just there and the open gate with a free passage into the road is somewhat indirect. The coalescence of the presentative and re-presentative elements into a conscious situation effective for the guidance of behaviour was not effected at once. After some ten or twelve experiences, in each of which the exit was more rapidly effected with less gazing out at wrong places, the fox terrier had learnt to go straight and without hesitation to the right spot. In this case the lifting of the latch was unquestionably hit on by accident, and the trick was only rendered habitual by repeated association in the same situation of the chance act and the happy escape. Once firmly established, however, the behaviour remained constant throughout the remainder of the dog's life, some five or six years.

Mr. E. J. Shellard observed * an act of similar import in a Scotch staghound, which "appeared at first to be the result

* "Introduction to Comparative Psychology," p. 290.

of thought," but which, on closer observation, was clearly seen to be the result of intelligence in the restricted sense of the

FIG. 21.—Fox-terrier lifting the latch of a gate.

term. The dog released the lever-latch of a yard door. "At first he raised his paws to the door and scratched violently, manifesting various signs of impatience. His scratches, which extended from the top of the door downwards, and over the whole area, would thus inevitably at some time or other reach the handle of the latch, which was thus struck forcibly downwards, the latch itself rising upwards. The door would then open from the weight of the dog pushing against it. The dog always opened the door in this manner from the time when the incident was first noticed until he left, a period of about three years. The door was opened with no greater ease at the expiration of that period than at the commencement. His paws would strike various parts of the door, and he never appeared to exercise any degree of judgment in the localization of his strokes, the fact of his paws striking the handle of the latch being a necessary result, providing the dog had sufficient patience and strength to continue."

One or two more experiments with my fox terrier may be briefly described. I watched his behaviour when a solid indiarubber ball was thrown towards a wall standing at right angles to its course. At first he followed it right up to the wall and then back as it rebounded. So long as it travelled with such velocity as to be only just ahead of him he pursued the same course. But when it was thrown more violently, so as to meet him on the rebound as he ran towards the wall, he learnt that he was thus able to seize it as it came towards him. And, profiting by the incidental experience thus gained, he acquired the habit—though for long with some uncertainty of reaction—of slowing off when the object of his pursuit reached the wall so as to await its rebound. Again, when the ball was thrown so as to glance at a wide angle from a surface, at first—when the velocity was such as to keep it just ahead of him—he followed its course. But when the velocity was increased he learnt to take a short cut along the third side of a triangle, so as to catch the object at some distance from the wall. A third series of experiments were made where a right angle was formed by the meeting of two surfaces. One side of the angle, the left, was dealt with for a day or two. At first the ball was directly followed. Then a short cut was taken to meet its deflected course. On the fourth day this method was well established. On the fifth, the ball was thrown so as to strike the other or right side of the angle, and thus be deflected in the opposite direction. The dog followed the old course (the short cut to the left) and was completely non-plussed, searching that side, then more widely, and not finding the ball for eleven minutes. On repeating the experiment thrice, similar results were that day obtained. On the following day the ball was thrown just ahead of him, so as to strike to the right of the angle, and was followed and caught. This course was pursued for three days, and he then learnt to take a short cut to the right. On the next day the ball was sent, as at first, to the left, and the dog was again non-plussed. I did not succeed in getting him to associate a given difference of initial direction with a resultant difference of deflection.

I may here mention that, whenever searching for a ball of which he had lost sight in the road, he would run along the gutter first on one side and then on the other. A friend who was walking with me one day regarded this as a clear case of rational inference. "The dog knows," he said, "the effects of the convex curvature of the road as well as we do." I am convinced, however (having watched his ways from a puppy), that this method of search was gradually established on a basis of practical experience. No logical inference on his part is necessary for the interpretation of the facts; and we should not assume its presence unless the evidence compels us to do so.

Dr. E. L. Thorndike, in a monograph on " Animal Intelligence " published as a supplement to the *Psychological Review* (June, 1898), has ·fully described and carefully discussed a number of interesting experiments. The subjects (one might, alas ! almost say victims) of some of these were thirteen kittens or cats from three to eighteen months old. His method of investigation shall be stated in his own words.

" After considerable preliminary observation of animals' behaviour under various conditions, I chose for my general method one which,

simple as it is, possesses several other marked advantages besides
those which accompany experiment of any sort. It was merely to
put animals when hungry in enclosures from which they could escape
by some simple act, such as pulling at a loop of cord, pressing a lever,
or stepping on a platform. The animal was put in the enclosure, food
was left outside in sight, and his actions observed. Besides recording
his general behaviour, special notice was taken of how he succeeded
in doing the necessary act (in case he did succeed), and a record
was kept of the time that he was in the box before performing the
successful pull, or clawing, or bite. This was repeated until the
animal had formed a perfect association between the sense-impression
of the interior of that box and the impulse leading to the successful
movement. When the association was thus perfect, the time taken
to escape was, of course, practically constant and very short.

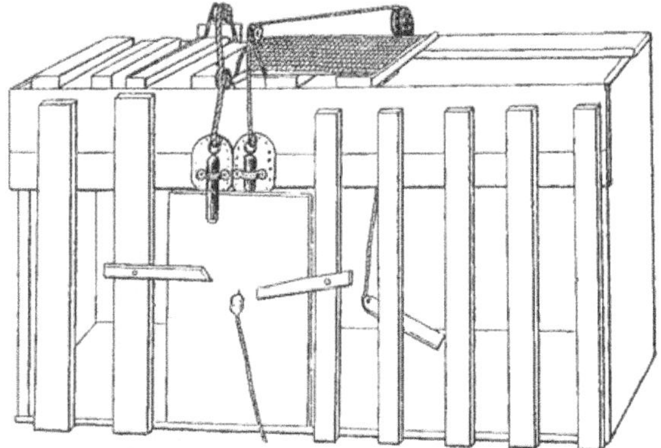

FIG. 22.—Cage used in Professor Thorndike's experiments.

"If, on the other hand, after a certain time the animal did not
succeed, he was taken out, but *not fed*. If, after a sufficient number
of trials, he failed to get out, the case was recorded as one of complete
failure. Enough different sorts of methods of escape were tried to
make it fairly sure that association in general, not association of a
particular sort of impulse, was being studied. Enough animals were
taken with each box or pen to make it sure that the results were not
due to individual peculiarities. None of the animals used had any
previous acquaintance with any of the mechanical contrivances by
which the doors were opened. So far as possible the animals were
kept in a uniform state of hunger, which was practically utter hunger."

To Dr. Thorndike's monograph we must refer those who desire detailed information as to apparatus and procedure. It must here suffice to state that the box-cages employed were rudely constructed of wooden laths, and formed cramped prisons about twenty inches long by fifteen broad and twelve high. Nine contained such simple mechanisms as Dr. Thorndike describes in the passage above quoted. When a loop or cord was pulled, a button turned, or a lever depressed, the door fell open. In another, pressure on the door as well as depression of a thumb-latch was required. In one cage two simple acts on the part of the kitten were necessary, pulling a cord and pushing aside a piece of board ; and in yet others three acts were requisite. In those boxes from which escape was more difficult a few of the cats failed to get out. The times occupied in thoroughly learning the trick of the box by those who were successful are plotted in a series of curves, the essential feature of which is the graphic expression of a gradual diminution in the time interval between imprisonment and escape in successive trials. This is shown in Fig. 23, which is constructed from some of Dr. Thorndike's data. In some cases the cats were set free from a box when they (1) licked themselves or (2) scratched themselves.

Dr. Thorndike comments on the results of his experiments as follows :—

" When put into the box the cat would show evident signs of discomfort and of an impulse to escape from confinement. It tries to squeeze through any opening; it claws and bites at the bars or wire; it thrusts its paws out through any opening, and claws at everything it reaches; it continues its efforts when it strikes anything loose and shaky: it may claw at things within the box. It does not pay very much attention to the food outside, but seems simply to strive instinctively to escape from confinement. The vigour with which it struggles is extraordinary. For eight or ten minutes it will claw, and bite, and squeeze incessantly. . . The cat that is clawing all over the box in her impulsive struggle will probably claw the string, or loop, or button so as to open the door. And gradually all the other nonsuccessful impulses will be stamped out, and the particular impulse leading to the successful act will be stamped in by the resulting pleasure, until, after many trials, the cat will, when put in the box, immediately claw the button or loop in a definite way. . . . Starting, then, with its store of instinctive impulses, the cat hits upon the

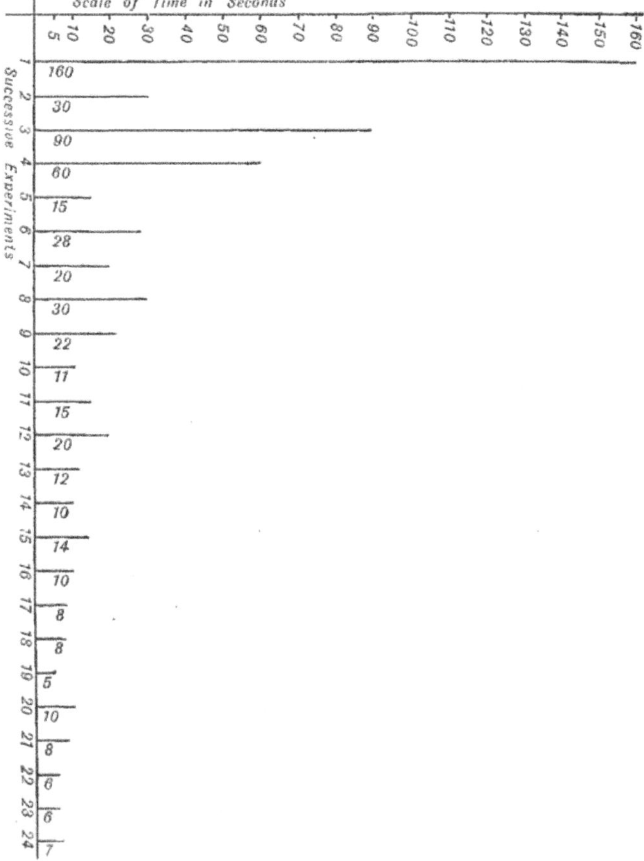

Fig. 23.—Diagram showing times taken by a kitten to escape from the cage in twenty-four successive experiments.

successful movement, and gradually associates it with the sense-impression of the interior of the box until the connection is perfect, so that it performs the act as soon as confronted with the sense-impression. . . . Previous experience makes a difference in the quickness with which the cat forms the associations. After getting out of six or eight boxes by different sorts of acts, the cat's general tendency to claw at loose objects within the box is strengthened and its tendency to squeeze through holes and bite bars is weakened; accordingly it will learn associations along the general line of the old more quickly. Associations between licking or scratching and escape are similarly established, and there was a noticeable tendency to diminish the act until it becomes a mere vestige of a lick or scratch. After the cat gets so that it performs the act soon after being put in, it begins to do it less and less vigorously. The licking degenerates into a mere quick turn of the head with one or two motions up and down with tongue extended. Instead of a hearty scratch, the cat waves its paw up and down rapidly for an instant."

Such experiments carried out on a different method give results in line with my own. The conditions are, however, somewhat unnatural, which I regard as in some respects a disadvantage. But we need experiments on different methods—the more the better,—and if the results they furnish are in accord, their correctness will be rendered the more probable. It is to be hoped that Dr. Thorndike will devise further experiments in which (1) the conditions shall be somewhat less strained and straitened, while the subjects are in a more normal state of equanimity (cannot "utter hunger" be avoided?), and (2) there shall be more opportunity for the exercise of rational judgment, supposing the faculty to exist. To establish the absence of foresight in the procedure of the cats, it is surely necessary so to arrange matters that the connections are clearly open—nay, even obvious—to the eye of reason. It appears that this consideration has not weighed sufficiently with Dr. Thorndike.

A series of experiments were made to ascertain whether instruction (in the form of putting the animal through the procedure requisite for a given act) was in any degree helpful. The conclusion is that such instruction has no influence. Those who have had experience in teaching animals to perform tricks will probably agree here—though some trainers

give expression to a different opinion.   It is, however, essential
to distinguish carefully between showing an animal how a trick
is done, and furnishing useful accessory stimuli (such as the
occasional taps of the trainer's whip when he wants a performing horse to kneel), or affording suitable conditions the
results of which temporarily enter into the association complex.   If the latter be eliminated the practice of trainers,
I believe, bears out the general result of the experiments.
Dr. Thorndike never succeeded in getting an animal to
change its way of doing a thing for his.   Nor was I, after
repeated trials, able to modify the way in which my dog
lifted the latch of the gate.   He did it with the back of his
head.   I could not get him to do it (more gracefully) with
his muzzle.

It may be said that the remarkable feats of performing
animals imply the existence of faculties of a higher order than
Dr. Thorndike and I are prepared to admit on the basis of our
experiments.   Mr. P. G. Hamerton many years ago described *
how, in his own house, a cleverly trained dog would fetch in
their right order the letters which spelt the English or German
equivalents of common French words, and do other wonderful
things.   But the owner of the dog (M. du Rouil) admitted
that there was a means of rapport between them which he was
not prepared to divulge.   It is just because the trainer has to
lead up to and utilize chance experiences that such prolonged
patience and care are required.   The animal is but the instrument on which his clever trainer plays ; an instrument of
wonderful intelligence, but lacking in the higher rational
faculty.   The organized scheme is the master's, not that of
his willing slave.   A rational being might not do more
wonderful things ; but he would learn them more rapidly
and by a less wearisome method.   As it is, the clever performing dog originates little or nothing, and repeats again and
again the same stereotyped behaviour, which—if one witnesses
the performance often—touches one with a profound sense of
its lack of rational spontaneity.

* *The Portfolio* (1873), p. 27, " Canine Guests."

As at present advised, therefore, I see no reason for withdrawing from the position provisionally taken up. The utilization of chance experience, without the framing and application of an organized scheme of knowledge, appears to be the predominant method of animal intelligence.

On this view, then, we may see in instinctive behaviour, and the multifarious automatic acts of animals, a means of providing experience of the right kind and on profitable lines. We may see in the play-instincts of the young a training ground for the more serious business of animal life—a theme developed by Professor Groos. We may see in the imitative tendency—the innate proclivity to follow a lead blindly and at first unintelligently—a further means of providing those useful items of experience which intelligence finds so serviceable. And we may see in the intelligence which can profit by chance occurrences that arise in these several ways all that suffices for the simple needs of animal existence.

With some differences of opinion Dr. Thorndike and I have much in common in the conclusions to which we have been independently led as to the method and limits of animal intelligence. We seem to be in essential agreement in the belief that the method of animal intelligence is to profit by chance experience without rational foresight, and that unless such experience be individually acquired, the data essential for intelligent progress are absent. While in our attempts to realize the general nature of animal consciousness there is a close similarity of treatment. In my " Introduction to Comparative Psychology " a good deal of space is devoted to an analysis of the psychology of skill " in order that we may infer what takes place in the minds of animals ; " and I said :—
" When I am playing a hard game of tennis, or when I am sailing a yacht close to the wind in a choppy sea, self does not at all tend to become focal. Hence, though I am a self-conscious being I am not always self-conscious. And presumably when I am least self-conscious, I am nearest the condition of the animal at the stage of mere sense experience. I am exhilarated with the sense of pleasurable existence, my

whole being tingles with sentient life. I sense, or am aware of, my own life and consciousness, in an unusually subtle manner. Experience is vivid and continuous. Such I take it to be the condition of the conscious but not yet self-conscious animal."

I can therefore cordially endorse Dr. Thorndike's conclusions as expressed in the following passages :—

"One who has watched the life of a cat or dog for a month or more under test conditions, gets, or fancies he gets, a fairly definite idea of what the intellectual [intelligent] life of a cat or dog feels like. It is most like what we feel when consciousness contains little thought about anything, when we feel the sense-impressions in their first intention, so to speak, when we feel our own body, and the impulses we give to it. Sometimes one gets this animal consciousness while in swimming, for example. One feels the water, the sky, the birds above, but with no thoughts about them or memories of how they looked at other times, or æsthetic judgments about their beauty ; one feels no ideas about what movements he will make, but feels himself make them, feels his body throughout. Self-consciousness dies away. Social consciousness dies away. The meanings, and values, and connections of things die away. One feels sense-impressions, has impulses, feels the movements he makes ; that is all."

And after an illustration from such a game as tennis, Dr. Thorndike adds : "Finally the elements of the associations are not isolated. No tennis-player's stream of thought is filled with free-floating representations of any of the tens of thousands of sense-impressions or movements he has seen and made on the tennis-court. Yet there is consciousness enough at the time, keen consciousness of the sense-impressions, impulses, feelings of one's bodily acts. So with the animals. There is consciousness enough, but of this kind."

It may be said that between the method of intelligence and that of fully developed rational procedure there is a wide gap which must have been bridged in the course of mental evolution. Unquestionably. And in contending that the methods of the animal are predominantly intelligent, I am far from wishing to assert dogmatically that in no animals are there even the beginnings of a rational scheme. Indications thereof do not indeed at present appear to have been clearly disclosed by experiment. But the experimental development of the subject is still in its infancy. We shall probably have to await the further results which must be the outcome of patient and well-directed child-study. The human child does pass in the course of his individual development from intelligent to rational procedure. Here there is a bridge which is crossed by every child. When we know more about the stadia of this development we shall be in a position to apply the results obtained in child-study in the analogous field of animal-study. Till then we must possess our souls in patience, and base our provisional conclusions on the results of systematic investigation, rather than on those of casual observation and anecdote.

IV.—The Evolution of Intelligent Behaviour

It is difficult to say where, in the hierarchy of animal progress, the beginnings of intelligence can first be traced. In the articulated animals, such as the insects, spiders, and crustacea, there is abundant evidence of intelligence of a relatively high grade. Many molluscs unquestionably profit by experience. The way in which limpets return to the scars on the rock which form their homes seems to show that they have acquired a practically adequate experience of their near surroundings. Romanes cites * some of the earlier observations which were extended by Professor Ainsworth Davis.† I looked into the matter myself some years ago, at Mewps Bay near Lulworth in Dorsetshire. The method adopted ‡ was to remove the limpets from the rock, and affix them at various distances from their scars. This can be done without difficulty or injury to the mollusc if one catches them as they are moving.

* " Animal Intelligence," pp. 28, 29.   † *Nature*, vol. xxxi., p. 200.
‡ *Ibid.*, vol. li., p. 127.

But one must make sure that they are just leaving or returning
to their proper homes, and are not taken in the midst of a
more extended peregrination, as in that case their special scars
cannot be noted. Failure to be careful in this matter vitiated
my earlier observations, which are therefore excluded in the
following table :—

Number removed.	Distance in inches.	Number returned.		
		In 2 tides.	In 4 tides.	Later.
25	6	21	—	—
21	12	13	5	—
21	18	10	6	2
36	24	1	1	3

From the nature of the rock surfaces the removal of a
limpet to a distance of two feet almost invariably involved
placing them on the further side of an angle. And though
some returned over such an angle, the majority did not.

In most cases the individuals which failed to return to
their respective scars took up new positions; and in several
instances, when they were subsequently removed to a distance
of a few inches from this new position, they returned to it.
Their return to the scar was watched in many cases, and the
course was fairly, but not quite direct. One limpet covered a
distance of ten inches, over a somewhat curved course, in a
little under twenty minutes. In another case the limpet on its
return journey had to pass between two others, which necessi-
tated the lifting of the shell to some height so as to clear one
of them. On reaching the scar they twist and turn about so
as to fit down in the normal position which is constant. When
they come up the wrong way round they rotate pretty rapidly
through the 180 degrees to get into position. One was
observed to make a short excursion from and return to its scar
under stillish water. But as a rule they seem to remain fixed
when they are submerged, moving for the most part when the
tide has just receded.

The greatest distance I have watched a limpet reach from its home was twenty-two inches.   But I have found them at a distance of three feet from their scars—that is to say, from those to which they fitted perfectly.   This was on a large flat surface.

When they move, the tentacles are projected out beyond the shell, and keep on touching and slightly adhering to the rock.   On reaching the scar they carefully feel round it with the tentacles.   By excision of these feelers Professor Davis was led to conclude that it is not through their instrumentality that the limpet finds its way back to its particular scar.   But I am inclined to question these results.   At any rate, further observations and experiments are needed to settle the point.

Snails will also return to special dark hollows or crannies in the wall after their foraging excursions.   Such behaviour in molluscs affords evidence of something more than instinct. In popular speech, we should say that there is memory of the locality.   And in any case it is difficult to interpret the facts without the assumption that the animals are conscious, and that re-presentative states are evoked through the mediation of presentative sense-impressions.   And such re-presentative states are the foundation-stones of experience, which forms the basis on which intelligent behaviour is grounded.

The most highly developed molluscs are the cephalopods. They have long sensitive mobile arms with which they feel for and capture their prey.   "Now Schneider observed," writes Dr. Stout,* "a very young octopus seize a hermit-crab.   The hermit-crab covers the shell in which it takes up its abode with stinging zoophytes.   Stung by these, the octopus immediately recoiled and let its prey escape.   Subsequently it was observed to avoid hermit-crabs.   Older animals of the same species managed cleverly to pull the crab out of its house without being stung."   Such cases afford evidence of profiting by experience through the exercise of intelligence.

Darwin's careful observations on the manner in which earthworms drag leaves into their burrows seem to show that

* " Manual of Psychology," p. 257.

these annelids act intelligently, and deal with leaves of different shapes in different ways. The leaves of Pine trees, consisting of two needles arising from a common base, were almost invariably drawn down by seizing this basal point of junction ; while the leaves of the Lime were, in 79 per cent. of the cases examined, drawn down by the apex ; in only 4 per cent. by the base ; and in the remaining 17 per cent. by seizing some intermediate portion. On the other hand, the leaves of the Rhododendron, in which the basal part of the blade is often narrower than the apical part, were in 66 per cent. of the observations drawn down by the narrower base. Triangles of paper were in the majority of cases seized by the apex. Commenting upon his observations, carried out with great care under experimental conditions, Darwin says,* " As worms are not guided by special instincts in each particular case, though possessing a general instinct to plug up their burrows, and as chance is excluded, the next most probable conclusion seems to be that they try in different ways to draw in objects, and at last succeed in some one way ; " that is to say, they profit by experience based on the method of trial and failure. But Darwin adds that the evidence he obtained shows " that worms do not habitually try to draw objects into their burrows in many different ways." And he seems to attribute to them an almost rational power of dealing with the circumstances in the light of general conceptions. " If worms," he says, " are able to judge, either before drawing or after having drawn an object close to the mouths of their burrows, how best to drag it in, they must acquire some notion of its general shape. This they probably acquire by touching it in many places with the anterior extremity of their bodies, which acts as a tactual organ. It may be well to remember how perfect the sense of touch becomes in a man when born blind and deaf, as are worms. If worms have the power of acquiring some notion, however rude, of the shape of an object and of their burrows, as seems to be the case, they deserve to be called intelligent ; for they

* " Vegetable Mould and Earthworms," p. 95.

then act in nearly the same manner as would a man under similar circumstances."

Such power of perceiving the relation of the shape of a leaf or other object to the form of the burrow is presumably beyond the reach of an earthworm. It may be regarded as more probable that the earthworm inherits an instinctive tendency to draw down objects in special ways, and that this is subject to some modification under the play of experience, without the formation of anything so psychologically complex as a general notion, however rude. In any case the behaviour of earthworms in closing their burrows seems to afford indications of something more than instinct—of that profiting by the results of experience which characterizes intelligent procedure.

Professor Whitman * has made some interesting observations on the leech *Clepsine*. " Place the animal," he says, " in a shallow, flat-bottomed dish, and leave it for a few hours or a day, in order to give it time to get accustomed to the place, and come to rest on the bottom. Then, taking the utmost care not to jar the dish or breathe upon the surface of the water, look at the *Clepsine* through a low magnifying lens, and see what happens when the surface of the water is touched with the point of a needle held vertically above the animal's back. If the experiment is properly carried out, it will be seen that the respiratory undulations (if such movements happen to be going on) suddenly cease, and that the animal slightly expands its body and hugs the glass. Wait a few moments until the animal, recovering its normal composure, again resumes its respiratory movements. Then let the needle descend through the water until the point rests on the bottom of the dish at a little distance from the edge of the body. Again the movements will cease, and the animal will hug the glass with its body somewhat expanded. Now push the needle slowly along towards the leech, and notice as the needle comes almost in contact with the thin margin of the body, that the part nearest the needle begins to retreat slowly before it.

* *Wood's Holl Biological Lectures* (1898), p. 287.

This behaviour shows a surprising keenness of tactile sensibility, the least touch of the water with a needle-point being felt at once. . . . If its back were rubbed with a brush or the handle of a dissecting needle, in order to test its sensitiveness to touch, the appearance would probably be that of insensibility and indifference to the treatment. Closer examination, however, would show that the flesh of the animal was more rigid than usual, and that the surface was covered with numerous stiff, conical elevations, the dermal papillæ or warts, which are so low and blunt in the normal state of rest as to be scarcely visible. It would be seen that the animal, although motionless, was in a state of active resistance to attack. . . . *Clepsine* has another and entirely different method of keeping quiet. The animal rolls itself up (head first and ventral side innermost) into a hard ball, outwardly passive, free to roll or fall whithersoever gravity or currents of water may direct it. . . . If by chance the animal has eggs, it will not desert them to escape in this way. . . . This species, then, has two quite distinct and peculiar ways of keeping quiet, and thus avoiding its enemies. If the animal has no eggs, or if it has young, it may adopt either mode of escape, while if it has eggs it has no choice but to remain quiet over them. . . . The act of rolling up into a passive ball may be performed (*a*) *under compulsion*, as when it is her last resort in self-defence; (*b*) *under a milder provocation*, as *one* of three courses of behaviour, as when the resting-place is turned up to light, and the choice is offered between remaining quiet in place, creeping away at leisure, or rolling into a ball and dropping to the bottom ; (*c*) or finally, *under no special external stimulus*, but rather *from internal* motive, the normal demand for rest and shady seclusion, presumably very strong in *Clepsine* after gorging itself with the blood of its turtle host."

Professor Whitman rightly regards the act of rolling into a ball as instinctive, and due to natural selection. But he does not undertake to discuss the question as to how much intelligence, if any, *Clepsine* may have. Nor, indeed, is it an easy matter to determine. The differential reaction according as the animal has eggs or not suggests intelligence ; but it may be instinct varying according to the conditions of stimulation external and internal.   The different behaviour which may be seen in different cases when a stone is turned to the light again suggests intelligence, but again may be determined directly by the conditions of stimulation.   Prompted by Dr. Whitman's observations, I endeavoured to determine whether a leech would grow accustomed to frequent gentle stimulation with a camel's-hair brush, and cease to react under circumstances which were followed by no ill effects.   But though I incline to think that this is the case, the observations were not such as to be satisfying and convincing.   If intelligence be present we seem to find it in an early and rudimentary state.

Observation and experiment seem to afford little indication of the conditions under which intelligence first makes its appearance in the animal kingdom.   And if we turn to general considerations, which at the best afford uncertain guidance, little light is thrown on the subject.   If we accept the view already indicated,* that the nerve-centres which are concerned in the conscious control begotten of experience are independent of those primarily concerned in normal reflex action, we may perhaps believe that the simplest nervous system, worthy of the name, contains both these elements, and that in the course of the evolution of nervous systems in higher and higher grades, there go on *pari passu* the further differentiation of these elements, and the progressive integration of reflex and control centres into a closely connected and effective whole.   Not that any expression of the facts, if such they be, in terms of an evolution formula, adds anything to our knowledge of the organic *modus operandi*.   We know but little of the intimate nerve physiology of even the highest invertebrates.   We see ample evidence of the control of behaviour in the light of individual experience.   Of any detailed knowledge concerning the manner in which this control is effected we do not seem to possess more than the rude initial phases.

* *Supra*, p. 44.

M

When we compare, however, the several grades of intelligence which observation discloses, and when we watch the conscious development of the more intelligent animals, we seem to find evidence of the growth of a system of experience, at first in very close touch with inherited modes of procedure, but gradually acquiring more of independence and freedom. Increase of the range and complexity of behaviour brings with it, not only increase in the range and complexity of experience, but also—what is, perhaps, even more essential to effective progress—greater unity and closer connection into a well-knit whole. And with this greater unity and closer connection there goes what one may term a condensation of experience by an elimination of detail and the survival of essential features repeatedly emphasized. This is analogous in the development of intelligence to the generalization and abstraction which play so important a part in the development of reason. It affords, in fact, the data which reflection utilizes in the purposive and intentional condensation and concentration of knowledge at a higher stage of mental development.

The omission of detail and the survival of the salient features is well known to us in the familiar facts of memory. We have seen thousands of sheep and oxen, no two of which are probably alike in all their external details as presented to vision. But we remember what a sheep or an ox looks like, and many of us can form a visualized image of either of these animals. This, however, is not the re-presentative image of any particular sheep or ox. It is what psychologists term a generic image. It is like a composite photograph made by superimposing on the same plate a number of individual images so that the salient features which all possess in common stand out clearly by their coincidence on the plate, while the distinctive details are but dimly presented. Thus does memory preserve the essentials common to many impressions while the distinguishing details are lost and fade, eliminated by forgetfulness. And thus in the experience which intelligence practically utilizes are the net results of a thousand particular impressions condensed in one effective image.

Condensation of experience is also effected by the elimination, under the guidance of consciousness, of those modes of behaviour which are not efficacious—a process to which Professor Mark Baldwin applies the phrase Functional Selection. There is a tendency at first to the overproduction of relatively useless actions. The multifarious random movements of the human infant, though their inexactness renders the child terribly helpless, afford a wide store of plastic material which intelligence can guide to its appropriate use. And the prolonged period of pupilage in the child is correlated with an unsurpassed range of combination and recombination of the abundant plastic material. The hereditary legacy, though it contains fewer drafts for definite and specific purposes than are placed to the credit of an animal rich in instinctive endowment, affords a far larger general fund on which intelligence may draw for the varied purposes of the freer financial existence of a rational being.

The relatively helpless young of many of the higher mammalia exhibit also much overproduction of seemingly aimless movements. But from these intelligence selects those which are of value for the purposes of life—those which experience proves to be effective. These—the relatively few—afford the motor impressions which by repetition stand out in experience, while the rest lapse from memory and are eliminated from experience as they are eliminated from practical performance. This is a great gain. Motor experience is rendered generic; the composite image that is retained is the net result of effective behaviour; and all that is valuable in the acquisitions of early life is condensed within manageable limits.

This process of rendering generic the particular items of a widening experience has a marked effect in the development of the conscious situations in the light of which behaviour is intelligently guided. It is not the master holding this whip or that ball which suggests to the dog a hiding or a scamper ; it is a generic situation with interchangeable details. It is not this, that, or the other previously unseen cat that at once

determines the situation for the fox terrier; the particular
animal has never entered into his past experience : it is the
fulfilment of the essential conditions of the generic image that
is operative in behaviour. The experience of animals must
inevitably become in large degree generic by the elimination
of the unessential and survival in re-presentative consciousness
of the salient elements in many slightly diverse situations.

Stated in terms of this conception, the familiar phenomena
of mimicry are due to the fact that the mimicking form accords
sufficiently well with the generic image to be taken as a
representative thereof. As is well known, the model has been
proved in many cases to be unpalatable or hurtful, while the
mimic is in itself neither the one nor the other. The drone-
fly, *Eristalis*, mimics the drone. And it has been urged that
this cannot be a true case of mimicry, since the drone is harmless, though the female and "neuter" bees are possessed of stings.
But I have satisfied myself by experiments with young birds, that
(1) after experience with bees drones are avoided, and (2) that
after similar experience drone-flies are also left untouched.
Hence it seems that all three fall within the same generic image,
the points of resemblance outweighing the differences in detail
—as they do, indeed, with many men and women.

Such examples of mimicry belong to what is known as the
" Batesian type "—so called after H. W. Bates, who, in 1861,
discussed its occurrence among Amazonian insects in the light
of the theory of natural selection. There are, however, certain
groups of insects which, although themselves " protected,"
possess common warning colours, causing them to resemble
each other. These are sometimes classed under the head of
" Mullerian mimicry "—so called after Fritz Müller, who, in
1879, first offered an explanation of the facts based on the
theory of natural selection. He suggested that such mutual
resemblance is advantageous to both protected forms, since it
lessens the number of those which are killed by young birds
and other animals while they are learning by experience what
to eat and what to leave. For, as the result of careful
observation, Mr. Frank Finn concludes " that each bird has

to separately acquire its experience, and well remembers what it has learnt,"—a conclusion with which, as already stated, my own observations are entirely in accord. There is therefore a certain amount of destruction of even well-protected forms by young and inexperienced birds. If, then, two such forms resemble each other, the acquisition of experience is thereby facilitated and the amount of destruction reduced, on the assumption that the two fall within the same generic image. Upholders of natural selection are not, indeed, at one in accepting this explanation, and further observation is unquestionably needed.   It is not improbable, however, that common protective coloration, such as the banding of yellow and black, seen in such different forms as the caterpillar of the cinnabar moth and the imago of the wasp, is of mutual utility. The following experiment was made with young chicks. Strips of orange and black paper were pasted beneath glass slips, and on them meal moistened with quinine was placed. On other plain slips meal moistened with water was provided. The young birds soon learnt to avoid the bitter meal, and then would not touch plain meal if it were offered on the banded slip. And these birds, save in two instances, refused to touch cinnabar caterpillars, which were new to their experience. They did not, like other birds, have to learn by particular trials that these caterpillars are unpleasant. Their experience had already been gained through the banded glass slips ; or so it seemed. I have also found that young birds who had learnt to avoid cinnabar caterpillars left wasps untouched. Such observations must be repeated and extended. But they seem to show that one aspect of the Müllerian theory is not without some facts in support of it ; and, so far as they go, they afford evidence that black and orange banding, irrespective of particular form, may constitute a guiding generic feature in the conscious situation.

It may be said that the generic condensation of experience here indicated implies the formation of general and abstract ideas, and that we cannot in face of the evidence accept Locke's dictum that abstraction is " an excellency which the faculties of brutes do by no means attain to." Romanes contended * that "all the higher animals have general ideas of 'good-for-eating' and 'not-good-for-eating' quite apart from any particular objects of which either of these qualities happens to be characteristic," and he quoted with approval Leroy's statement,† that a fox "will see snares when there are none; his imagination, distorted by fear, will produce deceptive shapes, to which he will attach an abstract notion of danger." According to such views animals form concepts; and concepts belong to the sphere of rational thought. It is not my intention to enter at length into the refinements of psychological distinction. Many psychologists, however, seek to distinguish between, on the one hand, the predominance by natural emphasis, of certain qualities, such as that of being suitable for food, and, on the other hand, the intentional isolation of these qualities for the purposes of thought and rational explanation. Abstraction they regard as a deliberate process applied with rational intent to the material afforded by experience and reflection. Generalization, too, they regard as deliberate, and carried out with like intent. The result is not merely a composite or generic product, but something more subtle and less dependent on sense. "All trees hitherto seen by me," said Noiré, "leave in my imagination a mixed image, a kind of ideal presentation of a tree. Quite different is my concept, which is never an image." The concept "tree" is a deliberate synthesis of abstract qualities intentionally isolated, and recombined in accordance with the general relationships which subsist between them.

If we accept this distinction, if we regard abstraction and generalization as intentional mental processes carried out with the rational intent of discovering the relationships of phenomena with the object of explaining them and recombining their essential features in an ideal scheme of thought, we shall probably admit, with John Locke, that these are excellencies which the faculties of brutes do by no means attain to. But we shall none the less see that the predominance of certain

* "Mental Evolution in Man," p. 27.
† "Intelligence of Animals," p. 121

salient features in experience by reiterated emphasis in association with natural needs, and the development of generic in place of merely particular re-presentations will afford the appropriate material for abstraction on the one hand, and generalization on the other. Intelligence supplies the embryonic mental structures from which, under the quickening influence of a rational purpose, abstract and general ideas may be evolved.

The essential features of the evolution of intelligence seem, then, to be, first, the development of controlling nerve-centres, by which the responsive action of reflex automatic or instinctive centres may be checked, augmented, or modified ; secondly, the increased differentiation and integration of these control centres with extension of the range and complexity of experience in close touch with practical needs ; thirdly, the condensation and concentration of experience by the formation of generic products through the reiterated emphasis begotten of recurrent situations having certain salient features in common, though differing in details ; and fourthly, an increased plasticity of behaviour, especially in early life, enabling an animal to deal effectually with an environment more complicated than that to which the more stereotyped instinctive behaviour is fitted by inheritance to respond. And this evolution of intelligent behaviour is working its way up to, though as such it cannot reach, the succeeding phase of mental evolution in which the data, supplied by intelligence, are treated with a new purpose for higher ends in the rational thought which seeks to explain the phenomena, and frame an ideal scheme of their relations and interconnections.

Two further points may be noticed. First, that it is during the early and plastic days or months of life that intelligence is setting its seal on animal behaviour, and stamping it with its distinctive character. Adult life is very much what youth has made it ; and old age is stereotyped through habit. In times of progress, the character of the race is determined by plastic possibilities of the young. Among them it is that the incidence of elimination makes itself felt, resulting in the survival of those whose intelligence can mould behaviour in accordance with the new circumstances of a wider life.

Secondly, this selection of the intelligent involves the survival of those in whose higher brain-centres there is room for a greater range and variety of interconnection by means of associating fibres. It involves a selective survival of the larger and more finely organized brains. It is probable, as Professor Ray Lankester has recently indicated, that the ridiculously small-brained mammals and reptiles of the past were creatures of instinct with little capacity for intelligent control. Their lives were simple, and their enemies and competitors no better provided with higher brain-centres than themselves. Stereotyped instinctive behaviour sufficed to enable them to hold their own, and meet the requirements of a life of dull and unprogressive monotony. Strength without cunning made these big-framed animals for a while masters of the situation. But there are no existing animals, whose skeletons indicate so high a position in the zoological series, which exhibit a cerebral development so poor. And we may fairly conclude that the fact that these huge creatures have left no lineal descendants may be taken as evidence of the importance and value, in evolution, of that cerebral tissue which is the organic basis of intelligence. The higher brain contains the potentiality of that experience without which the evolution of intelligent behaviour in any race of vertebrate animals is impossible.

V.—The Influence of Intelligence on Instinct

We have seen that the relation of instinct to intelligence is essentially that of congenital to acquired behaviour. We have seen, too, that in the Lamarckian interpretation what is acquired in the course of life may be transmitted through inheritance, and thus the intelligent behaviour of one generation may become instinctive and congenital in the next. But serious biological difficulties stand in the way of the acceptance of this interpretation ; there is, moreover, little or no evidence of the assumed transmission to offspring of any acquired modifications of structure or behaviour. We have, therefore, been led to infer that instinctive behaviour has been evolved through the selection of adaptive variations of germinal origin, the influence of intelligence being restricted to the fosterage of co-incident variations, that is to say, of those congenital variations which coincide in direction with the acquired modifications of behaviour due to intelligence. It is clear that on this interpretation the influence of intelligence on instinct is more indirect and less simple than that implied by the Lamarckian hypothesis. Intelligence and instinct are in large degree independent, though there is continual interaction between them. We have now to consider the nature of this interaction, and to this end we must indicate the relation of acquired modifications to the hereditary groundwork of the animal constitution.

The basal fact is, that the bodily tissues are subject to a certain amount of structural change during the course of individual life in accordance with the amount of functional strain put upon them. The labourer's thickened skin, the enlarged and strengthened muscles of the athlete, the juggler's acquired suppleness are familiar cases. Less familiar instances are afforded under abnormal conditions. Should one kidney from any cause be slowly destroyed, the other will slowly enlarge to carry on the increased work of elimination of waste products ; when the larger shin bone of a dog has been removed after injury, the smaller bone becomes thickened to bear the added strain ; new joint surfaces are sometimes formed where bones have been broken and the natural joints injured.

One may say that the normal development of any structure depends upon a due amount of use. But, since in the course of strenuous life any organ is from time to time subject to an abnormal amount of strain, it must be fitted to respond to a super-normal call on its strength and functional activity. Were the heart and the lungs, for example, unable to meet the greatly added drain on their energies, due to unwonted and severe exertion, collapse, perhaps death, would ensue if such exertion were imperatively demanded under special circumstances. And it is clear that many wild animals must be not infrequently placed in such circumstances as will subject their muscular structures and the functional activity of their organs of circulation and respiration to a strain nearly up to their extreme limits of endurance. The carnivorous hunter would often fail to secure his prey if his organization were unequal to a hard and prolonged chase; the hunted prey would not survive to procreate his kind if he fell a victim to the first pursuer through inability to stand the exertion necessary to enable him to make good his escape. It is thus, we may believe, through natural selection that a sufficiently high standard of strength and functional endurance is maintained. The failures in these respects are steadily eliminated. It is difficult to realize the great strain put upon a bird's organization by the migration flight. Some ten times as many birds leave our shores in the autumn as return to them in the following spring. What proportion of these is weeded out in the act of migration we do not know; but we may be sure that only those fitted to stand a severe test of physical endurance return to rear broods which shall inherit in large degree similar vigour of constitution.

Two factors, then, determine the limits of efficiency in the bodily organs—heredity and use. And these two co-operate in such a way that we may say, either that due use is the essential condition of the effective development of the hereditary powers, or that heredity serves to condition their effective development through use. But though closely related, so that each may be regarded as conditional on the other, they are, if we accept the view that acquired characters are not transmitted as such, so far independent in that use adds nothing to, disuse subtracts nothing from, the hereditary store. It is, indeed, difficult to conceive how, on any view, the absence of the conditioning factor of normal use can be the efficient cause of a positive diminution of the balance at the bank of heredity. And Lamarckian thinkers have not succeeded in placing their conception of the matter in the clear light of a working hypothesis.

The amount of what we may term " modifiability " by use differs a good deal in the several organs and tissues. The teeth of carnivora and the antlers of deer may be cited as structures in which the conditioning effects of use form a relatively unimportant factor. On the other hand, the nervous system, with which we are here primarily concerned, is of all animal structures that in which what is acquired may attain the greatest importance in the successful conduct of life ; the nature and the range of behaviour affording an index of the amount of modifiability in this respect.

We have already seen that instinctive behaviour is primarily a matter of the first occasion on which any given action is performed, and that many instinctive acts are subject to subsequent modification in the light of the experience gained during the early performances. The range of such modification varies both in different animals and also with respect to different modes of behaviour in the same animal. The more fixed and deeply rooted an instinct the less readily does intelligence obtain a hold on it, so as to direct the behaviour into new channels of better accommodation to the circumstances. M. Fabre describes how a *Sphex*, one of the solitary wasps, instinctively draws its prey, a grasshopper, into the burrow by its antennæ. When these were cut off the wasp pulled the grasshopper in by the jaw appendages ; but when these were removed she seemed incapable of further accommodation to the unusual circumstances. It would seem an easy and obvious application of intelligence to seize the prey by one of the forelegs. But this was not done ; and the grasshopper was then left. Intelligence did not seem equal to meeting the altered conditions presented by the maimed grasshopper. Still, there was some modification of the normal instinctive behaviour ; and, as Dr. Peckham has shown, there may be more than Fabre noted. Let us assume the existence of an animal whose every act is instinctive, whose whole behaviour is marked out in strictly hereditary lines, no

new departures being acquired in the course of individual life.
This extreme case would afford an example of what we may
term completely stereotyped behaviour.   On the other hand, let
us assume the existence of an animal with no hereditary
definiteness of reaction, whose every act is intelligent, whose
whole behaviour is the result of individual acquisition.   This
antithetical extreme case would afford an example of what we
may term completely plastic behaviour.   It is questionable,
however, whether either of these extreme types occur in nature.
What we find in our study of animal behaviour is some intermediate condition in which both factors co-operate, with a
predominance either of stereotyped instinctive response on the
one hand, or of plastic intelligent acquisition on the other
hand.   And in the latter case, as such behaviour approaches its
ideal limits, we have modifiability under the circumstances of
individual life at its maximum.

The evolution of intelligence as such runs parallel with the
evolution of plastic behaviour ; and this plasticity is necessi-
tated by the variety and the complexity of the conditions of
life—a variety and a complexity requiring many subtle modifications of response to enable the behaviour to reach accommodation to the changeful exigencies of diverse circumstances.   To
meet constant and relatively fixed conditions stereotyped
instinctive responses suffice ; and the elimination under natural
selection of those individuals which fail to respond in fixed
ways by specially adaptive behaviour tends to render definite
the hereditary channels of nervous intercommunication.   An
inherited system of no little complexity may thus be evolved ;
of which we have seen examples in our study of instinctive
behaviour.   But the essential condition of the successful
working of such a system is unvarying constancy in the
environment to which the stereotyped instinctive behaviour is
adapted.   Instinct without intelligence is like a barrel-organ
constructed to play a limited number of tunes with monotonous
precision.   The music of its behaviour depends entirely on
what the maker, heredity, has inserted in the works.   But
though a barrel-organ may suffice where a hymn-tune, a jig, a

hornpipe, and a funeral march exhaust all the possible requirements, it is sadly lacking in musical plasticity. To obtain that, you must place intelligence at the keyboard that the music may be accommodated to a greater number of varying moods.

It may be urged that such an illustration is, in many respects, obviously faulty ; and that barrel-organs do not under any circumstances develop into musicians. No doubt the illustration is faulty. But it may be questioned whether instinct under any circumstances develops into intelligence, any more than a barrel-organ into a musician. As we said at the beginning of this section, intelligence and instinct are in large degree independent though there is continual interaction between them. Completely stereotyped behaviour, in its theoretical perfection, is in exact adaptation to the circumstances. Where instincts are only relatively perfect, further adaptation is secured through congenital variation and the survival of the individuals in which the behaviour is better adapted to the comparatively invariable circumstances. This is one line of evolution. But it no more contains within itself the potentiality of developing into plastic accommodation to varying circumstances, than the barrel-organ contains within itself the potentiality of becoming a musician. The evolution of intelligence is along independent lines of progress. Stereotyped adaptation can never pass up into plastic accommodation. These two belong to independent lines of evolution ; but they are bound up in the same nervous system, they jointly determine the behaviour, they interact not only in the course of individual life but in the process of evolution, and they are both subject to the incidence of natural selection, which can determine whether the one line or the other shall preponderate—whether instinct or intelligence shall dominate behaviour.

If an answer must be given to the question whether instinct or intelligence has priority in the course of the evolution of behaviour, it may be urged that, on theoretical grounds, the claims of instinct are the stronger. No doubt the evolution of the two lines of development have proceeded to a large extent side by side ; but whereas intelligence does in a number

of cases demonstrably modify the course of instinctive behaviour,
the converse proposition does not hold good in the same sense.
The animal acts at first instinctively, and subsequently in the
light of experience reaches further accommodation ; and though
in later life a dominating instinct may override the guidance
of intelligence, still, even this is probably due to the fact that
the instinct is more deeply seated in the constitution than any
opposing habits of intelligent acquisition.

We can, however, infer what is the influence of intelligence
on instinct without basing our inferences on any assumption
of the initial priority of instinct in the evolutionary sequence.
Taking animals as we find them, they afford numberless examples of behaviour at first instinctive but subsequently modified, in greater or less degree, in accordance with the teachings
of experience. Let us, first, assume that the environment is
slowly changing, or has changed, in some definite manner.
Such change would, of course, be relative, and might be due,
either to new conditions brought to bear on the animal, or to
the animal being itself brought, in the expansion of its life,
within their influence. The old instinct is no longer quite
adapted to the changed circumstances. If the change were
sufficient in amount, and occurred somewhat suddenly, variations
of instinct might not occur soon enough to enable the animal
to reach adaptation by the gradual process of natural selection.
If dependent on instinct alone the animal would, under these
circumstances, be eliminated. But if intelligence were able to
modify the behaviour to meet the new conditions this elimination would be prevented. In successive generations intelligence
would constantly modify behaviour in the same manner and in
a definite direction. Meanwhile congenital variations in
different directions would occur. Those which were in directions antagonistic to that dictated by intelligence would tend
to thwart accommodation and render it less effectual ; but
those which were coincident in direction would conspire with
accommodation and render it more effectual. The individuals
in which variations of instinct tended to thwart intelligence
would be eliminated ; while those in which coincident variations

assisted and aided intelligent modification would survive. Thus intelligence would lead the way along lines which congenital variations would follow. And in the course of a number of generations the new instinct would reach the fully adaptive level, and further modification by intelligence would become unnecessary unless the environment continued to change yet more. Individual accommodation of behaviour would in this way determine the direction of instinctive variation; and yet throughout the process there would be, strictly speaking, no transmission of the intelligently acquired characters of the behaviour.

But though under constant and uniform changes in the environment the net result would be only a guided variation of the original instinct, under more variable and indefinitely changing circumstances the result would be different. The higher animals exhibit an intelligent plasticity which enables them to meet the requirements of the more complex environment into which their wider life has risen ; for evolution lifts the animal from narrower into progressively wider spheres of activity and behaviour, so that its environment becomes relatively more complex. Here stereotyped behaviour would be rather a hindrance than an advantage. The winning animal in life's struggle would be the one in which behaviour was most rapidly and most surely modified to meet particular needs—the one in which the teachings of experience were most promptly utilized in effective action. The inevitable tendency of the evolution of intelligence must be disintegration of the stereotyped modes of behaviour and the dissolution of instinct. Natural selection, which under a uniform and constant environment leads to the survival of relatively fixed and definite modes of response, under an environment presenting a wide range of possibilities leads to the survival of plastic accommodation through intelligence. It is not that intelligence has any direct influence tending to undermine the hereditary foundations of instinct, for acquired plasticity is not inherited as such ; it is rather that when the stereotyped and the plastic are pitted against each other in the struggle for existence in

the wider, freer, and more varied life of the higher animals
the plastic survives and the stereotyped succumbs.

Imperfect as is our present knowledge of the manner
in which the nervous connections implied in psychological
associations are established, there can be no question that
they are acquired in the course of individual life ; they are
modifications of nervous structure due to a special mode of
use under the conditions of experience. Here, then, in the
case of the nervous system, as in that of the bodily organs
before mentioned, two factors determine the limits of efficiency
—heredity and use. And these two, again, co-operate in such
a way that we may say, either that due use—here represented
by adequate experience—is the essential condition of the
effective development of the hereditary potentialities, or that
heredity serves to condition their effective development
through use and experience. And just as the heart and lungs
must inherit the power of standing abnormal strain if the
animal is to avoid elimination in times of unwonted exertion,
so must the nervous system inherit some reserve power of
dealing effectively with unwonted circumstances by intelligent
accommodation, if the animal is not to fall a victim to such
circumstances. In other words, at times of heightened competition those animals which can draw on a reserve fund of
intelligent accommodation will survive, while the stupid
blunderers will be eliminated. We may term this reserve
fund of intelligent accommodation, this inherited ability to
meet specially difficult circumstances as they arise, *innate
capacity*. From the nature of the case it must be indefinite,
for it must carry with it the ability to meet unforeseen
combinations of the environing forces by new combinations
of the results of experience. Its distinguishing mark is
plasticity, in contradistinction to the stereotyped fixity of
typical instinct. And accompanying its evolution there is
probably, as we have seen, a dissolution of its antithesis,
instinct. Thus may we account for the fact that man, with
his great store of innate capacity, has so small a number of
stereotyped instincts.

But the dissolution of instincts is not complete. Residua are left in the inherited mental constitution. And these we term *congenital tendencies* and *propensities*. They differ from the typical instincts in the fact that the definiteness of response has been lost. They dictate a general trend of action, but the particular application in behaviour is due to intelligent accommodation. They are commonly spoken of as instinctive ; and their mode of origin justifies the use of the adjective in association with the term "propensities." But it must be remembered that the behaviour to which they lead is not, as such, wholly instinctive ; it is a joint product of instinct and intelligence, the general trend being due to the instinctive propensity, while the mode of application is guided by intelligence.

There is, however, another way in which analogous propensities may be ingrained in the mental constitution. It is a well-known and familiar fact that the frequent repetition of intelligent accommodation in certain definite lines begets habits, which so far simulate instincts as to be commonly described in popular speech as instinctive. Professor Wundt indeed places them in the category of "acquired instincts "— a usage which we regard as unsatisfactory, seeing that it tends to mask the distinction between the congenital and acquired factors in behaviour, and seeing that we have the well-defined term "habits" for acts rendered to a large extent automatic through repetition. Lamarckian thinkers regard habit as the mother of instinct, assuming that the acquired automatism of one generation may be transmitted to become congenital in the succeeding generation. This conclusion we provisionally reject regarding the basal assumption as at present unproven. But though we cannot accept the view that habit is the mother of instinct, we regard it as not improbable that habit may be the nurse of congenital propensities. Remembering that similar habits are acquired by animals of the same species throughout a series of succeeding generations, and assuming that congenital variations are constantly occurring in many directions, it seems probable that some of these variations will be coincident in direction with the acquired habits. Thus would arise a congenital propensity to perform the habitual acts ; and should they be of sufficient importance in the conduct of life to be subject to the action of natural selection, those animals in which such propensities were congenital would survive, whereas those in which no such propensities existed would be eliminated. It is unnecessary, however, to elaborate this conception further, since it is in line with that already discussed in considering the influence of intelligence in fostering a diversion of instinct under changing circumstances.

Sufficient has now been said to illustrate some of the ways in which instinct and intelligence interact in the evolution of behaviour. Such interaction is further exemplified in the social life of animals, which will be dealt with in the next chapter.

CHAPTER V

SOCIAL BEHAVIOUR

I.—Imitation

The characteristic feature of social behaviour is that it is in large degree determined by the behaviour of other members of the social community. In all animals which mate there is a temporary or more lasting influence on each other of the individuals which unite to procreate their kind ; and in those which foster their young there is a social relation of parents and offspring. Some of these mutual relationships will be discussed, in their emotional aspects, in the next chapter. Here we will consider the more general factors which serve to determine the course of social evolution.

Among these is commonly reckoned imitation. M. Tarde says, " La société c'est l'imitation." But this word, like so many others which are employed alike in popular speech and in more or less technical discussions, carries a somewhat wide range of meaning, and is by some writers used in a broader, by others in a narrower sense. Thus Professor Mark Baldwin * says, " that all organic adaptation in a changing environment is a phenomenon of *biological* or *organic imitation*," under which category will fall, therefore, the organic behaviour of the protozoa and of plants. On the other hand, Professor E. L. Thorndike, though he admits in the lower animals " certain pseudo-imitative or semi-imitative phenomena," has been led by experiments, to be presently noticed, to the conclusion that

* " Mental Development in the Child and the Race—Methods and Processes," p. 278.

animals as high in the scale of life as cats and dogs cannot
form new associations under the influence of imitation. "It
seems sure," he says,* "from these experiments, that the
animals were unable to form an association leading to an act
from having seen another animal, or animals, perform the act
in a certain situation." In face of such apparently diverse
usage it is necessary to show within what limits and with
what qualifications the word may profitably here be used to
indicate a factor in social evolution.

Professor Mark Baldwin's use of the term "imitation" can
only be understood in its relation to an hypothesis of organic
and mental evolution, which he develops with no little skill
and brilliancy.† He regards the processes of life as issuing in
a great twofold adaptation, due to expansions and contractions,
—the former representing waxing, the latter waning vitality;
and he holds that all special adaptations are secured by the new
hold upon beneficial stimulations reached by the expansive out-
reaching movements. "Among the variations in organic
forms," he says, "it is easy to see that some of them might
react in such a way as to keep in contact with the stimulus, to
lay hold of it, and so keep on reacting to it again and again—
just as our rhythmic action in breathing keeps the organism in
vital contact with the oxygen of the air. These organisms will
get all the benefit or damage of the repetition or persistence of
the stimulus, or of their own reactions, again and again ; and
it is self-evident that the beneficial stimulations are the ones
which should be maintained in this way, and that the organisms
which did this would live. The organisms which reacted in
such a way as to retain the damaging stimulations, on the other
hand, by this same process, would aid nature in killing themselves. If this be true, only those organisms would survive
which had the variation of retaining useful stimulations in
what I have called, in speaking of imitation elsewhere, a ' circular
way ' of reacting. . . . So, when we come to consider phylogeny

* "Animal Intelligence :" monograph supplement to *Psychological
Review*, 1898, p. 61.

† *Op. cit.*, pp. 263, 172, 201, 132, and 248 (note).

and ontogeny together, we find that if by an organism we mean a thing of contractility or irritability, whose round of movements is kept up by some kind of nutritive process supplied by the environment — absorption, chemical action of atmospheric oxygen, etc.—and whose existence is threatened by dangers of contact and what not, the first thing to do is to secure a regular supply to the nutritive processes, and to avoid these contacts. But the organism can do nothing but move, as a whole or in some of its parts. So, then, if one of such creatures is to be fitter than another to survive, it must be the creature which, by its movements, secures more nutritive processes and avoids more dangerous contacts. But movements toward the source of stimulation keep hold on the stimulation, and movements away from the contacts break the contacts ; that is all. Nature selects these organisms; how could she do otherwise ? "

"Thus a 'circular' activity is found in operation ; life-processes issuing in increased movements, by which in turn the stimulations to the life-processes are kept in action." But when a child imitates, himself reproducing the " copy " set for imitation, the reaction at which imitative suggestion aims is one which will *reproduce the stimulating impression*, and so tend to perpetuate itself. The stimulus starts a motor process, which tends to reproduce the stimulus, and, through it, the motor process again. It is a "circular activity." Thus "we are able to reconstruct the theory of adaptation in such a way as to show that this kind of organic selection by movement, and this kind of imitative selection by consciousness, are the same thing. Organic imitation and conscious imitation—each a circular process tending to maintain certain stimulations and to avoid others—here is one thing ;" and to this one thing the common term "imitation " is applied by Mr. Baldwin.

This extended usage is admitted by the author to be somewhat of an innovation. But if his hypothesis be sound this need be no bar to its acceptance. Two salient questions must, however, receive satisfactory answers. First, is all organic adaptation in a changing environment a circular process—a phenomenon of organic imitation? Secondly, does all conscious imitation tend to reproduce the imitating stimulus?

Professor Baldwin speaks of organic imitation and conscious imitation as "each a circular process tending to maintain certain stimulations and to avoid others." Now, it may be granted that the tendency to maintain or repeat certain stimulations may be regarded as a "circular process." But can the avoidance or non-repetition of others be so regarded? A large proportion alike of the hereditary adaptations and the acquired accommodations of behaviour are directed to this avoidance or non-repetition of hurtful stimulations. The instinctive shrinking of a chick from an aggressive animal is just as much adaptive as the repeated cuddling beneath the warm wing of the mother. The avoidance of nauseous cinnabar caterpillars is just as much an accommodation to the constitution of the environment as the reiterated seizing of palatable grubs. Even low down in the scale of animal life, Dr. Jennings's observations on Paramecia seem to show that the retention of favourable stimulation is not due to its direct influence, but is the indirect result of a reaction to the relatively unfavourable stimulation which occurs when the Paramecium passes away from more satisfactory surroundings. A favourable environment is secured through the avoidance of the unfavourable. Unless, therefore, we exclude adaptive avoidance from the category of adaptations, we cannot regard all organic adaptation in a changing environment as a phenomenon of organic imitation due to a circular process tending to the reinstatement of stimulation.

Passing to the second question—Does all conscious imitation tend to reproduce the initiating stimulus?—we cannot unreservedly give an affirmative answer. It is true that when a child more or less successfully reproduces a sound which falls upon its ear, a like sound stimulus is afforded which may by a circular process incite to renewed effort, and lead to yet more successful reproduction. But when Professor Baldwin's child, between nine and ten months old, imitated certain movements of the lips, there was no reproduction of the initiating visual stimulus. A chick seeing its companions run away or crouch will follow suit ; and this would commonly be termed an imitative action ; but there is here no reproduction of the initiating stimulus. Very much of the behaviour which is usually ascribed to imitation produces effects in consciousness quite different from that of the suggestive stimulation. It is only by selecting one's examples that one finds in them evidence in favour of Professor Baldwin's "circular process."

Since, therefore, this circular mode of activity is neither a characteristic of all conscious imitation, nor a distinguishing mark of all adaptive organic action, the grounds on which Professor Baldwin bases his extended usage of the term appear to be fallacious. And in this usage we cannot follow him.

Turning now to Professor Thorndike's very different contention—that animals even so high as the cat and dog do not imitate in the sense of forming an association leading to an act from having seen another animal perform the act in a certain way—we may first describe some of his ingenious experiments designed to submit the matter to the test of observation under controlled conditions.[*]

Experiments were made with chicks in several ways. They were, for example, placed in pens, from which, in each case, "there was only one possible way of escape, to see if they would learn it more quickly when another chick did the thing several times before their eyes. The method was to give some chicks their first trial with an imitation possibility, and their second without, while others were given their first trial without and their second with. If the ratio of the average time of the first trial to the average time of the second is smaller in the first class than it is in the second class, we may find evidence of this sort of influence by imitation. Though imitation may not be able to make an animal do what he would otherwise not do, it may make him do *quicker* a thing he would have done sooner or later anyway. As a fact, the ratio is *much longer*. This is due to the fact that a chick, when in a pen with another chick, is not afflicted by the discomfort of loneliness, and so does not try to get out. So the

* "Animal Intelligence," pp. 47–64.

other chick, who is continually being put in with him to teach him the way out, really prolongs his stay in. This factor destroys the value of these quantitative experiments, and I do not," says Mr. Thorndike, "insist upon them as evidence against imitation, though they certainly offer none for it."

Chicks, from sixteen to thirty days old, were also placed in boxes from which escape was open to them by such acts as pecking at the door, stepping on a platform, or pecking at a tack. The method of experiment was to put a chick in, leave him from sixty to eighty seconds, then put in another who knew the act, and on his performing it to let both escape. No cases were counted unless the imitator apparently saw the other do the thing. After about every ten such chances to learn the act, the imitator was left in alone for ten minutes. Out of thirteen cases tabulated only once was the act performed, in spite of the ample chance for imitation. "I have no hesitation," adds Mr. Thorndike, "in declaring this one's act in stepping on the platform the result of mere accident, and am sure that any one who had watched the experiments would agree."

To test the influence, if any, of imitation in cats, the following method was adopted. A box was arranged with two compartments separated by a wire screen. "The larger of these had a front of wooden bars with a door which fell open when a string stretched across the top was bitten or clawed down. The smaller was closed by boards on three sides and by the wire screen on the fourth. Through the screen a cat within could see the one to be imitated pull the string, go out through the door thus opened, and eat the fish outside. When put in this compartment, the top being covered by a large box, a cat soon gave up efforts to claw through the screen, quieted down, and watched more or less the proceedings going on in the other compartment. Thus this apparatus could be used to test the power of imitation. A cat who had no experience with the means of escape from the large compartment was put in the closed one; another cat, who would do it readily, was allowed to go through the performance of pulling the string, going out, and eating the fish. Record was made of the number of times he did so, and of the number of times the imitator had his eyes clearly fixed on him. . . . After the imitatee had done the thing a number of times, the other was put in the big compartment alone, and the time it took him before pulling the string was noted and his general behaviour closely observed. If he failed in five or ten or fifteen minutes to do so, he was released and not fed. This entire experiment was repeated a number of times. From the times taken by the imitator to escape and from observation of the way that he did it, we can decide whether imitation played any part. . . . No one, I am sure, who had seen the behaviour of the cats would have claimed that their conduct was at all influenced by what they had seen. When they did hit the string the act looked just like the accidental success of the ordinary association experiment. But, besides these personal observations, we have in the impersonal time-records sufficient proofs of the absence of imitation." Some observations on dogs are also described. From these it appears that the three individuals on which experiments were made failed to learn the way of getting out of a cage from seeing another dog escape. One of them was also allowed to see another dog beg for meat 110 times. But he never tried to imitate him and thus secure a piece of meat as a reward. It therefore " seems sure," says Mr. Thorndike, "that we should give up imitation as an *a priori* explanation of any novel intelligent performance. To say that a dog who opens a gate, for instance, need not have reasoned it out if he had seen another dog do the same thing, is to offer instead of one false explanation another equally false. Imitation in any form is too doubtful a factor to be presupposed without evidence."

Professor Thorndike is of opinion that monkeys are probably imitative in ways beyond the capacity of dogs and cats ; but, at the time of writing, he had not substantiated his opinion, by analogous experiments. If so, it will perhaps prove that they are rational beings in the narrower sense defined in a previous chapter of this work. For it appears that the kind of imitation which Mr. Thorndike's experiments go far to disprove, is what we may term reflective imitation. A cat with no experience of the means of escape, one that has tried to get out of the box by chance efforts in many directions and has failed, sees another cat perform an act acquired in this way, and learns nothing from the sight. This, no doubt, proves that the cat had not in any sense grasped the nature of the problem before it, had no notion of just where the difficulty lay, had not the wit to see that the performance of the other cat supplied the missing links in its own previous behaviour. It is questionable whether such missing links could be supplied in this way in the absence of some powers of reflection. The cat is unable to form an association, leading to an appropriate act, from having seen another animal perform the act in a certain way, partly because it cannot perceive the reason of its previous failure, and see that the other's performance affords the requisite clue. The whole gist of the chance experience interpretation of animal behaviour is that there must be chance experience to build on. The cat cannot gain this by looking on never so intently, unless it be provided with a rational as well as a sensory eye. The act of pulling the string has been reached by the successful cat through the gradual elimination of many failures ; it is a differentiated act, having no place in the previous experience of the kitten. It has never entered into the conscious situation, and cannot be supplied at will by a non-rational being.

As Mr. Thorndike himself says, " no cat can form an association leading to an act unless there is included in the association an *impulse* of its own which leads to the act." * By " impulse," Mr. Thorndike " means the consciousness accompanying a muscular innervation apart from that feeling of the act which comes from seeing one's self move, from feeling one's body in a different position, etc. It is the direct feeling of doing as distinguished from the idea of the act done gained through eye, etc. . . The act in this respect of being felt as to be done or as doing is in animals the

* *Op. cit.*, pp. 66, 14, 15.

important thing, is the thing which gets associated, while the act as done, as viewed from outside, is a secondary affair." I take it that by "impulse" is here meant what Dr. Stout would term the direct experience involved in conation.[*] If it have a place in experience distinguishable from that of stimulation and response it is included in what I have on a former page spoken of as the consciousness of behaviour as such, which was said to be essential. And I am surprised that Mr. Thorndike should have supposed that I believe that this could by any animal be "supplied at will." In any case it seems probable, as the result of observation, that unless the consciousness of behaving in a specific manner has entered into the situation as developed in experience it cannot in animals enter into any subsequent representative complex. And it is the absence of such consciousness of behaving in a specific manner which the sight of the escaping cat fails to supply in Mr. Thorndike's experiments.

Interesting and valuable as these experiments are, they are open to the criticism to which, as we have seen, his other experiments are also open—that the conditions are abnormal and cramped. Apart from reflective imitation, which they tend to disprove, they do not conduce to the kind of conscious situation which appears to be most favourable for the development of intelligent imitation founded on hereditary tendencies and propensities. It is through such imitation that, as Herr Groos says,[†] "animals learn perfectly those things for which they have imperfect hereditary dispositions." The kind of situation which conduces to such intelligent imitation is that which involves the attitude of attention and interest rising, when these are sufficiently varied in their direction, into what is spoken of as curiosity. These, in their natural occurrence in animals, are parts of, or in any case accompaniments of the conative attitude—they are connected with activities and impulsive tendencies to behaviour. If attention and interest are directed to the behaviour of another animal, the

---

[*] Cf. *infra*, p. 235.

[†] "The Play of Animals," Eng. trans., p. 79.

conative attitude is that of imitation. Miss Romanes has described how skilfully a capuchin imitated the actions necessary to unlock a trunk. It does not seem necessary to assume that reflective imitation is here exemplified. The monkey need not regard the key and lock as the related parts of a puzzle to be practically solved, need not have any free idea of the difficulty it presents, need not in unlocking the trunk grasp the true nature of the difficulty or have any conception of its solution. Every several act of the capuchin, the seizing the key, the directing it here or there, and so on, is already supplied with the impulse of which Dr. Thorndike speaks. Attention, itself charged with impulse, directs and combines these pre-existing impulses to a new end. And since that which directs the attention is the act of another, we call the procedure imitative. But the varied and persistent effort differs in no essential respect from that of a two days' chick, which pecks again and again at some speck which catches its eye, or that of a nestling jay, which will peck for long at some rail or piece of wire in its cage, twisting and turning its bill in many and varied ways. And success in opening the trunk is reached by the capuchin, not, it would seem, through any real appreciation of the essential kernel of the practical problem, but through the chance results of many varied efforts. Although in no other animals is it developed to so high a degree as in the monkeys, interest in the doings of others is an attitude by no means rare, and affords the basis of intelligent imitation. Perhaps the conditions in Dr. Thorndike's experiments were not the best for the development of such interest in the procedure of another. And in any case the imitation of a particular mode of procedure, reached by the gradual defining of the impulse, could hardly be expected in the absence of the series of experiences by which that definition had been reached, unless the cat were capable of what has been above spoken of as reflective imitation.

If, then, we agree to exclude from the category of imitative behaviour in animals, on the one hand, any " circular process " which may occur in the same individual, and on the other hand any reflective imitation, such as is so important a factor in human education, it remains to be seen what may be fairly included in this category.

It is probable that in animals imitation has its foundations in instinctive behaviour, of which it may be regarded as the characteristically social type. If one of a group of chicks learn by casual experience to drink from a tin of water, others will run up and peck at the water, and thus learn to drink. A hen teaches her little ones to pick up grain or other food by pecking on the ground and dropping suitable materials before them, while they seemingly imitate her action in seizing the grain. One may make chicks and pheasants peck by simulating the action of a hen with a pencil point or pair of fine forceps. According to Mr. Peal, the Assamese find that young jungle pheasants will perish if their pecking responses are not thus stimulated ; and Professor Claypole tells me that this is also the case with young ostriches hatched in an incubator. A little pheasant and guinea-fowl followed two older ducklings, one wild, the other tame, and seemed to wait upon their bills, to peck when they pecked, and to be guided by their actions. It is certainly much easier to bring up young birds if older birds are setting an example of eating and drinking ; and instinctive acts, such as scratching the ground, are performed earlier if imitation be not excluded. If a group of chicks have learnt to avoid cinnabar caterpillars, and if then two or three from another group are introduced and begin to pick up the caterpillars, the others will sometimes again seize them, though they would otherwise have left them untouched. One of my chicks, coming upon a dead bee, gave the danger or alarm note ; another at some little distance at once made the same sound. A number of similar cases might be given ; but what impresses the observer as he watches the early development of a brood of young birds, is the presence of an imitative tendency which is exemplified in many little ways not easy to describe in detail. It is probable, however, that these imitative tendencies or propensities are not wholly indefinite. The young birds do not imitate any actions, but behaviour of certain specific types, the imitation of which has been engrained through the action of natural selection.

What generalization, then, can be drawn from this somewhat indefinite group of facts, to which many others of like import could be added from observations on the young of mammals ? What is their relation to instinctive procedure in general ? It would seem that they are characterized by a special relation of the external stimulus to the response. When this stimulus is afforded by the behaviour of another animal, and the responsive behaviour it initiates is similar to that which affords the stimulus, such behaviour may be termed imitative. A chick sounds the danger note ; this is the stimulus under which another chick sounds a similar note, and we say that the one imitates the other. Such an action may be described as imitative in its effects, but not imitative in its purpose. Only from the observer's standpoint does such instinctive behaviour differ from other modes of congenital procedure. It may be termed biological, but not psychological, imitation. And if it be held that the essence of imitation lies in the purpose so to imitate, we must find some other term under which to describe the facts. This does not seem necessary, however, if we are careful to qualify the term " imitation " by the adjective " instinctive " or " biological." And the retention of the term serves to indicate that this is the stock on which deliberate imitation is eventually grafted.

The fact that instinctive imitation leads, under natural conditions, to behaviour which is already familiar to us in the species concerned, prevents us from recognizing the influence of this social factor so easily as might otherwise be the case. The abnormal arrests our attention more readily than the normal, and hence the cases commonly cited are generally those which strike us as unusual, such as the imitation of human sounds by the parrot. But if the young inherit a tendency to imitate certain actions of their parents, and if there is among the members of a gregarious species such instinctive imitation as shall tend to keep them gregarious, we have here a social factor in animal life of no slight importance.   Just as the higher type of reflective imitation is of great value in bringing the human child to the level of the adults who form the family and social environment, so, too, does the sub-conscious instinctive imitation of the lower animals bring the young bird or other creature into line with the members of its own species.   In broods of chicks brought up under experimental conditions, there are often one or two more active, vigorous, intelligent, and mischievous birds.   These are the leaders of the brood ; the others are their imitators.   Their presence raises the general level of intelligent activity.   Remove them, and the others show a less active, less inquisitive, less adventurous life.   They seem to lack initiative.   From which one may infer that imitation affords to some extent a means of levelling up the less intelligent to the standard of the more intelligent ; and of supplying a stimulus to the development of habits which would otherwise be lacking.   When a mongrel pup, whose development Dr. Wesley Mills watched and has described, was introduced to the society of other dogs, its progress was, he tells us, " extraordinarily rapid."

Instinctive imitation thus introduces into the conscious situation certain modes of behaviour, and if the development of the situation as a whole is pleasurable, there will be a tendency to its redevelopment, under the guidance of intelligence, on subsequent occasions.   As in the case of other instincts and propensities, there is given through inheritance a more or less definite outline sketch of social procedure, which intelligence further defines, and refines, and shapes to more delicate issues.   As a rule, however, intelligence does not tend to make the imitation as such more perfect.   It may perfect the behaviour, but not necessarily on imitative lines. In the case, however, of the song and call-notes of birds, and not improbably the sounds of other animals, there does seem a predisposition to render the imitation as such more perfect. The facts, as afforded by such birds as the magpie, jay, starling, marsh-warbler, and mocking-bird, are familiar ; and

I have elsewhere * given some account of them. It may be specially noted that we have in this case that circular mode of activity on which, as we have seen, Professor Mark Baldwin lays so much stress. Professor Thorndike seems to regard the phenomena presented by imitative birds as somewhat of a mystery, and as the result of a specialization removed from the general course of mental development. And he says that, until we know whether there is in birds which repeat sounds any tendency to imitate in other lines, we cannot connect these phenomena with anything found in the mammals, or use them to advantage in a discussion of animal imitation as the forerunner of human. Upon the view, however, that such imitation is primarily instinctive and only secondarily intelligent, there seems no reason why we should expect to find imitation in birds running along any other lines than those which the hereditary instinct has marked out. And so far from being unable to use the phenomena to advantage in a discussion of animal imitation as a forerunner of human, we may perhaps see in them the best examples, other than those afforded by apes, of that intelligent imitation which is the precursor of the rational and reflective imitation of the boy or girl.

In the case of the human child we may see the three stages in the development of imitation. First, the instinctive stage, where the sound which falls upon the ear is a stimulus to the motor-mechanism of sound production. Secondly, the intelligent stage of the profiting by chance experience. Intelligence, as we have seen, aims at the reinstatement of pleasurable situations, and the suppression of those which are the reverse. The sound-stimulus, the motor effects in behaviour, and the resulting sound-production coalesce into a conscious situation, which appears to be pleasurable or the reverse, according as the sound produced resembles or not the initiating sound-stimulus. If we assume that the resemblance of the sounds he utters to the sounds he hears is itself a source of pleasurable satisfaction (and this certainly seems to be the case), intelligence,

* " Habit and Instinct," pp. 174-180.

without the aid of any higher faculty, will secure accommodation and render imitation more and more perfect. And this appears to be the stage reached by the mocking-bird or the parrot. But the child soon goes further. He reflects upon the results he has reached ; he at first dimly, and then more clearly realizes that they are imitative ; and his later efforts at imitation are no longer subject to the chance occurrence of happy results, but are based on a scheme of behaviour which is taking form in his mind, are deliberate and intentional, and are directed to a special end more or less clearly perceived as such. He no longer imitates like a parrot ; he begins to imitate like a man, and may, by the study of good models and the maintenance of a high ideal, acquire the moving cadences of an orator.

According to our interpretation, instinctive imitation is a factor of wide importance in animal behaviour, intelligent imitation, arising in close connection with interest in the doings of others, is a co-operating factor, but of intentional and reflective imitation there is at present no satisfactory evidence in any animal below man.

II.—Intercommunication

The foundations of intercommunication, like those of imitation, are laid in certain instinctive modes of response, which are stimulated by the acts of other animals of the same social group. These have been fostered by natural selection as a means of social linkage furthering the preservation, both of the individual and of the group.

Some account has already been given of the sounds made by young birds, which seem to be instinctive and to afford an index of the emotional state at the time of utterance. That in many cases they serve to evoke a like emotional state and correlated expressive behaviour in other birds of the same brood cannot be questioned. The alarm note of a chick will place its companions on the alert ; and the harsh "krek" of a young moor-hen, uttered in a peculiar crouching attitude, will often throw others into this attitude, though the maker of the warning sound may be invisible. That the cries of her brood influence the conduct of the hen is a matter of familiar observation ; and that her danger signal causes them at once to crouch or run to her for protection is not less familiar. No one who has watched a cat with her kittens, or a sheep with her lambs, can doubt that such " dumb animals " are influenced in their behaviour by suggestive sounds. The important questions are, how they originate, what is their value, and how far such intercommunication—if such we may call it—extends.

There can be but little question that in all cases of animals under natural conditions such behaviour has an instinctive basis. Though the effect may be to establish a means of communication, such is not their conscious purpose at the outset. They are presumably congenital and hereditary modes of emotional expression which serve to evoke responsive behaviour in another animal—the reciprocal action being generally in its primary origin between mate and mate, between parent and offspring, or between members of the same family group. And it is this reciprocal action which constitutes it a factor in social evolution. Its chief interest in connection with the subject of behaviour lies in the fact that it shows the instinctive foundations on which intelligent and eventually rational modes of intercommunication are built up. For instinctive as the sounds are at the outset, by entering into the conscious situation and taking their part in the association-complex of experience, they become factors in the social life as modified and directed by intelligence. To their original instinctive value as the outcome of stimuli, and as themselves affording stimuli to responsive behaviour, is added a value for consciousness in so far as they enter into those guiding situations by which intelligent behaviour is determined. And if they also serve to evoke, in the reciprocating members of the social group, similar or allied emotional states, there is thus added a further social bond, inasmuch as there are thus laid the foundations of sympathy.

"What makes the old sow grunt and the piggies sing and whine?" said a little girl to a portly substantial farmer. "I suppose they does it for company, my dear," was the simple and cautious reply. So far as appearances went, that farmer looked as guiltless of theories as man could be. And yet he gave terse expression to what may perhaps be regarded as the most satisfactory hypothesis as to the primary purpose of animal sounds. They are a means by which each indicates to others the fact of his comforting presence ; and they still, to a large extent, retain their primary function. The chirping of grasshoppers, the song of the cicada, the piping of frogs in the pool, the bleating of lambs at the hour of dusk, the lowing of contented cattle, the call-notes of the migrating host of birds—all these, whatever else they may be, are the reassuring social links of sound, the grateful signs of kindred presence. Arising thus in close relation to the primitive feelings of social sympathy, they would naturally be called into play with special force and suggestiveness at times of strong emotional excitement, and the earliest differentiations would, we may well believe, be determined along lines of emotional expression. Thus would originate mating cries, male and female after their kind ; and parental cries more or less differentiated into those of parent and offspring, the deeper note of the ewe differing little save in pitch and timbre from the bleating of her lamb, while the cluck of the hen differs widely from the peeping note of the chick in down. Thus, too, would arise the notes of anger and combat, of fear and distress, of alarm and warning. If we call these the instinctive language of emotional expression, we must remember that such "language" differs markedly from the "language" of which the sentence is the recognized unit.

It is, however, not improbable that, through association in the conscious situation, sounds, having their origin in emotional expression and evoking in others like emotional states, may acquire a new value in suggesting, for example, the presence of particular enemies. An example will best serve to indicate my meaning. "In the early dawn of a grey morning," says Mr.

H. B. Medlicott,* "I was geologizing along the base of the Muhair Hills in South Behar, when all of a sudden there was a stampede of many pigs from the fringe of the jungle, with porcine shrieks of *sauve-qui-peut* significance. After a short run in the open they took to the jungle again, and in a few minutes there was another uproar, but different in sound and in action ; there was a rush, presumably of the fighting members, to the spot where the row began, and after some seconds a large leopard sprang from the midst of the scuffle. In a few bounds he was in the open, and stood looking back, licking his chaps. The pigs did not break cover, but continued on their way. They were returning to their lair after a night's feeding on the plain, several families having combined for mutual protection ; while the beasts of prey were evidently waiting for the occasion. I was alone, and, though armed, I did not care to beat up the ground to see if in either case a kill had been effected. The numerous herd covered a considerable space, and the scrub was thick. The prompt concerted action must in each case have been started by the special cry. I imagine that the first assailant was a tiger, and the case was at once known to be hopeless, the cry prompting instant flight, while in the second case the cry was for defence. It can scarcely be doubted that in the first case each adult pig had a vision of a tiger, and in the second of a leopard or some minor foe."

If we accept Mr. Medlicott's interpretation as in the main correct, we have in this case : (1) common action in social behaviour, (2) community of emotional state, and (3) the suggestion of natural enemies not unfamiliar in the experience of the herd. Under uniform conditions of experience the alarm-notes of some birds may well call up, re-presentatively, salient features in previous situations. Unquestionably, in the parrot, the word-sounds they imitate become associated with definite objects of sense-experience. In the following case, a particular

* "The Evolution of Mind in Man," footnote, pp. 25, 26. Quoted in "Introduction to Comparative Psychology," from which the comments on it are extracted.

sound appeared to be suggestive of a particular sense-idea in the dog. The parent blackbirds, which built near a house in Clifton, were wont to give the alarm-note when marauding cats appeared in sight. This sound, it would seem, became definitely associated, in the experience of a terrier, with the animals the presence of which called it forth ; and on hearing the alarm note the dog would rush out into the garden, apparently, as I am informed by his mistress, in fullest expectation of a pleasant worry. It is a not improbable hypothesis, therefore, that in the course of evolution the initial value of uttered sounds is emotional ; but that on this may be grafted in further development the indication of particular enemies. If, for example, the cry which prompts instant flight among the pigs is called forth by a tiger, it is reasonable to suppose that this cry would give rise to a representative generic image of that animal having its influence on the conscious situation. But if the second cry, for defence, was prompted sometimes by a leopard and sometimes by some other minor foe, then this cry would not give rise to a re-presentative image of the same definiteness. Whether animals have the power of intentionally differentiating the sounds they make to indicate different objects, is extremely doubtful. Can a dog bark in different tones to indicate "cat" or "rat," as the case may be ? Probably not. It may, however, be asked why, if a pig may squeak differently, and thus, perhaps, incidently indicate on the one hand "tiger" and on the other hand "leopard," should not a dog bark differently, and thus indicate appropriately "cat" or "rat" ? Because it is assumed that the two different cries in the pig are the instinctive expression of two different emotional states, and Mr. Medlicott could distinguish them ; whereas, in the case of the dog, we can distinguish no difference between his barking in the one case and the other, nor do the emotional states appear to be differentiated. Of course, there may be differences which we have failed to detect. What may be regarded, however, as improbable, is the *intentional* differentiation of sounds by barking in different tones with the *purpose* of indicating "cat" or "rat." Mr. R. L. Garner, in a work * which unfortunately contains much hasty and immature generalization, distinguished nine sounds made by capuchins. But none of these, so far as can be gathered from the data given, is necessarily indicative of a particular object. All of them may be emotional expressions of satisfaction, discontent, alarm, apprehension, and so forth. In any case, there is no evidence for that intentional employment of sounds, to the realized end of intercommunication, which would involve the exercise of an incipient rational faculty. Such powers of intercommunication as animals possess are based on direct association, and refer to the here and the now. A dog may be able to suggest to his companion the fact that he has descried a worriable cat ; but can a dog tell his neighbour of the delightful worry he enjoyed the day before yesterday in the garden where the man with the biscuit-tin lives ? Probably not, bark he never so expressively.

Although some anecdotes are commonly interpreted as affording evidence of descriptive intercommunication among animals, we need the decisive results of experiment before this view can be unreservedly accepted. Sir John Lubbock, now Lord Avebury, made careful experiments with ants, and discusses the question with his customary lucidity and impartiality. "Much of what has been said," he writes,† "as to the powers of communication possessed by bees and ants depends on the fact that if one of them in the course of her rambles has discovered a supply of food, a number of others soon find their way to the store. This, however, does not necessarily imply any power of describing localities. If the ants merely follow a more fortunate companion, or if they hunt her by scent, the matter is comparatively simple ; if, on the contrary, the others have the route described to them, the case becomes very different." Experiments were therefore made to decide the question. For example, when an ant returned from the discovered store of food to the nest, and then emerged with a

* "The Speech of Monkeys."
† "Scientific Lectures," pp. 112, 118.

following of other ants, she was taken up on a slip of paper and transferred to the food. The followers, thus deprived of their leader, in nearly all cases failed to find the store. "I conclude, then," says Lord Avebury, "that when large numbers of ants come to food they follow one another, being also, to a large extent, guided by scent. The fact, therefore, does not imply any considerable power of intercommunication." There are, moreover, some circumstances which seem to strengthen this conclusion. For instance, "if a number of slave-ants are put in a box, and if in one corner a dark place of retreat be provided for them, with some earth, one soon finds her way to it. She then comes out again, and going up to one of the others, takes her by the jaws and carries her to the place of shelter. They then both repeat the same manœuvre with other ants, and so on until all their companions are collected together. Now, it seems difficult to imagine that so slow a course would be adopted, if they possessed any power of communicating description."

Lord Avebury is, however, of opinion that such insects can transmit simpler ideas. He found, for example, that where ants were put to a large and a small store of larvæ under similar circumstances, a greater number of insects followed the ant that had discovered the larger store. This may, indeed, have been due rather to a difference in manner than to any intentional communication; but the fact remains that through some difference of behaviour there resulted suggestive effects on other members of the community.

But although there can be little doubt that the behaviour of social insects has suggestive value for others, it may still be regarded as very doubtful whether they are able to communicate information to one another by any system of language or signs, purposively employed as a system to this end. The distinguished geologist, Hague, communicated to Darwin * the effects on ants of crushing some of their number as they proceeded along a definite trail. "As soon as those ants which were approaching arrived near to where their fellows lay dead

* *Nature*, vol. vii., p. 443.

and suffering, they turned and fled with all possible haste."
" When such an ant, returning in fright, met another approaching, the two would always communicate, but each would
pursue its own way, the second ant continuing its journey to
the spot where the first had turned about, and then following
that example." There seems nothing to show that the
" communication " here was effective.

From the many anecdotes of dogs calling others to their
assistance, or bringing others to those who feed them or treat
them kindly, we may indeed infer the existence of a social
tendency and of the suggestive effects of behaviour, but we
cannot derive conclusive evidence of anything like descriptive
communication. And although domestic animals may learn
or be taught to associate the words we utter with certain acts
or things, or may even, in a sense, communicate their wishes
to us by special modes of behaviour—as in the case of Lord
Avebury's poodle, Van,* who was taught to bring cards on which
such words as " Food " or " Out " were printed, and in that of
a cat which touches the handle of the door when she wants
it opened for her,—still, all these are founded on direct
association, and are in a line with the act of Mr. Thorndike's
cat, which licked herself or scratched herself when imprisoned
in a cage, such act having entered into the association-complex.

Such intentional communication as is to be found in animals,
if indeed we may properly so call it, seems to arise by an
association of the performance of some act in a conscious
situation involving further behaviour for its complete development. Thus the cat which touches the handle of the door
when it wishes to leave the room has had experience in which
the performance of this act has coalesced with a specific
development of the conscious situation. The case is similar
when your dog drops a ball or stick at your feet, wishing you
to throw it for him to fetch. And on these lines may probably
be interpreted such behaviour as Romanes † thus described :—
" Terrier A being asleep in my house, and terrier B lying on a

* " The Senses of Animals," p. 277.
† " Mental Evolution in Man," p. 100.

wall outside, a strange dog, C, ran along below the wall on the
public road, following a dog-cart. Immediately on seeing C,
B jumped off the wall, ran upstairs to where A was asleep,
woke him up by poking him with his nose in a determined
and suggestive manner, which A at once understood as a sign :
he jumped over the wall and pursued the dog C, although C
was by that time far out of sight round a bend in the road."
Romanes did not probably intend to imply that A by poking
B, conveyed specific information that there was another dog,
C, which had proceeded in a particular direction. That would
be descriptive communication. The meaning attaching to A's
action was presumably similar to that which characterizes
other " meaning " for intelligent animals—the development of
the situation on lines marked out by previous experience.
Still, it is clear that such an act would be the perceptual
precursor of the deliberate conduct of the rational being by
whom the sign is definitely realized as a sign, the intentional
meaning of which is distinctly present to thought. This
involves a judgment concerning the sign as an object of
thought ; and this is probably beyond the capacity of the dog.
For, as Romanes himself says,* " it is because the human
mind is able, so to speak, to stand outside of itself, and thus
to constitute its own ideas the subject-matter of its own
thought, that it is capable of judgment, whether in the act of
conception or in that of predication. We have no evidence
to show that any animal is capable of objectifying its own
ideas ; and, therefore, we have no evidence that any animal is
capable of judgment."

It seems, therefore, that the sounds made by animals, and
certain other modes of behaviour, may be regarded as primarily
instinctive acts which have been evolved with the biological
end of affording suggestive stimuli furthering intercommunication between the members of the social group. Their
performance, however, affords data to consciousness, which
intelligence makes use of in the guidance of behaviour in
accordance with the results of experience. And since the

* *Op. cit.*, p. 175.

similar acts performed by the socially linked members are in many cases closely connected with emotional states, there arises the further social link of community of feeling—that which, perhaps, more than anything else conduces to community of action and similarity of social behaviour. Occasionally particular sounds or special acts may, through constant and uniform association, indicate particular objects, such as natural enemies. But there does not appear to be convincing evidence of any intentional differentiation of the means of communication, or of any use of sounds for descriptive ends.

Still, just as the instinctive imitation we considered in the last section may be regarded as the precursor, in the animal world, of the reflective and rational imitation of which we may watch the development in children, so may instinctive modes of intercommunication be regarded as supplying the foundations on which deliberate and intentional communication may be based. And here imitation will be a co-operating factor. We see in the early stages of the development of children's language how large a share simple and direct association takes in the process. For a while, indeed, there seems to be this and nothing more. But gradually there arises a realization of a further import and purpose in the hitherto isolated associations. It is seen that they symbolize elements in that incipiently rational scheme of thought and things which is beginning to take form in the child's mind. The relationships which hold good within the conscious situations of daily life begin to occupy the focus of attention, and hitherto unappreciated word-sounds are perceived to stand out as signs for these relationships. Of course the relationships * are implicit in the conscious situations of the higher animals and of infants. Only by reflection can they become explicit, and rivet the attention. Something is needed to bring them into prominence and focus the mental eye upon them. And descriptive intercommunication supplies this need. If a description, even the simplest, is to be apprehended or presented to the apprehension of others, then

* Compare chap. xiii., on "The Perception of Relations," in my "Introduction to Comparative Psychology."

the relationships must be rendered explicit. Try to describe
an ordinary visual scene, or the most commonplace sequence
of events, and see if you can do so without making clear to the
mind the relationships involved. The thing is impossible. An
infant or a dog cannot understand the simplest possible
description, because the words and suffixes which indicate the
relationships have no meaning. The words which stand for
substantive impressions may have suggestive value through
direct association. The word " cat " or " rats " may have for
the dog a very definite suggestive value ; and hence some
people fancy that when they say to their dog, " There is a cat
in the garden," the animal understands what they say. But it
is quite sufficient to suppose that the word " cat " has suggestive
force, all the rest being for the dog mere surplusage of sound.
When we talk to our four-footed companions, how much can
they be said to understand of what we say ? Perhaps a score
of words have for a dog a definitely suggestive value, each
associated with some simple object or action. " Out," " down,"
" up," " walk," " biscuit," " cat," " fetch," and so forth elicit
appropriate responses. Even with these, tone is more suggestive than articulation, and in each word the salient feature
is the chief guide. When I said " Whisky," for example, to
my fox-terrier, he would at once sit up and beg ; not because
his tastes were as depraved as those of his master, but because
the *isk* sound, common both to " whisky " and " biscuit," was
what had for his ears the suggestive value.

In a paper on the " Speech of Children," * Mr. S. S.
Buckman exhibits the animal stage in the incipient speech of
the human infant. We cannot here discuss, still less criticize,
his paper. One or two examples will serve to illustrate how
instinctive sounds may serve as the basis for subsequent
speech. He regards *ma* as primarily a forcible expression of
an emotional state. " If the child require attention it makes
the loudest noise which it can produce ; the parting of the
lips and opening of the mouth to the widest extent while the
full volume of breath is emitted produces the sound *ma*." At

* *Nineteenth Century*, May, 1897, pp. 793–807.

first the sound seems to have the value of a simple expression of an emotional state. " But if the infant require attention it is its mother whom it wants, and from whom it receives this attention ; therefore *ma* very soon comes to be recognized as the call for mother, and, by a further step in development, as the name for mother." Here, if we accept the interpretation, we have the passage from the emission of a sound as the expression of emotion to the use of the sound from its association with a particular object of sense-experience to indicate that object. Similarly, according to Mr. Buckman, with *kah*. At first " a strong sign of displeasure at anything nasty to the taste," it passes, we are told, into a symbol for the bad ; hence κακός ; and is perhaps narrowed down to the particularly offensive κάκκη. *Da* and *ta* are regarded as recognition sounds, the former being associated eventually with the father, the latter with strangers. This appears somewhat hypothetical, but, granting the accuracy of Mr. Buckman's interpretation, these sounds also illustrate again the transition from the expression of an emotion to sounds indicative of particular objects of experience.

Interesting, however, as are such observations on the animal stage of sound-production in the human infant, they do not touch the crucial period in the development of language. Mr. Buckman, indeed, regards as a remarkably dogmatic assertion Professor Max Müller's dictum that " the one great barrier between the brute and man is language ; " and he tells us that " there are more than twelve different words in the language of fowls," on which assertion, in turn, the distinguished linguist whom he criticizes might have something piquant to say. No doubt the difference of opinion turns on the definition of the word " language." But if, as is now generally accepted, the sentence and not the word is the distinguishing unit in language, and the copula in some form, explicit or implicit, is the pivot of the sentence, the wisest hen is probably incapable of language. The word becomes an element in language—a word proper—only when it assumes the office of a part of speech, that is to say, a constituent element in an interrelated whole. The animal "word," if we like so to term it, is an isolated brick ; a dozen, or even a couple of hundred such bricks do not constitute a building. Language, properly so called, is the builded structure, be it a palace or only a cottage ; hen language, or monkey language, is, at best, so far as we at present have evidence, an unfashioned heap of bricks. It is just because language is the expression of a portion of a scheme of thought that it indicates in the speaker the possession of a rational soul, capable of perceiving and symbolizing the relationships of things as reflected in thought.

Herein lies the practical value, for human advance in mental development, of language as a means of descriptive intercommunication. It renders explicit relationships otherwise merely implicit, and forces them to the front ; and since these relationships are the stuff of which knowledge is built— without the realization of which any complex ideal scheme is impossible of attainment—the importance of descriptive intercommunication can scarcely be overestimated. And though there is no conclusive evidence of its occurrence among animals, yet we have in them the instinctive and intelligent basis on which in due course of evolution it may be securely based.

III.—Social Communities of Bees and Ants

Apart from human societies the most noteworthy social communities of animals are found among insects, especially in ants, bees, wasps, and termites. It is true that in the mammalia we find such communities as the troop of apes, the herd of cattle, the pack of wolves, the school of porpoises, the so-called "rookeries" of seals, and the colonies of "prairie dogs" and of beavers ; and that among birds there are analogous communities. Undoubtedly the temporary or permanent association of many individuals is in such cases an advantage to the race, and confers mutual benefits on the associates. But in none of these cases is division of labour carried to such a high degree as among the social insects.

And it is through such division of labour that the social community reaches its highest expression.

It is a somewhat remarkable fact that in man, where we find the social division of labour brought to a high pitch of perfection, and carried out with great nicety of accommodation to those circumstances which civilization has rendered extremely complicated, there is no organic differentiation of structure among the co-operating individuals; whereas, so low down in the scale of life as the colonial polype, *Hydractinia*, which is often found growing on the shells occupied by hermit crabs, there are at least three kinds of differentiated individuals : nutritive polypes with mouth and tentacles ; mouthless sensitive members ; and others whose sole office is reproduction. But these differentiated individuals in the colonial zoophytes are connected at their bases by a common flesh ; and the division of labour is a product of organic evolution, and is probably not in any degree determined or guided by consciousness. We may say, then, that the division of labour in the zoophyte is wholly physical, whereas in man it is chiefly conscious or psychical ; as is also the bond of union between the several members of the colony. Intermediate between these extremes stand the social insects. In them there is no physical bond of union, for each individual is distinct and separate ; the social linkage is in some degree conscious under the conditions of their nurture ; and the division of labour is partly conscious, though probably in large degree based on instinctive foundations, and partly the outcome of an organic differentiation of structure seen in the reproductive members and in the sterile workers, as exemplified in the common wood ant (Fig. 24). In some cases the workers themselves may be divided into different castes.

So much has been written—and well written—on the social life of insect communities, that it will here suffice to indicate some of the problems which arise when we endeavour to interpret the modes of behaviour which have been carefully observed. In the honey-bee we have the well-known differentiation of structure into drones or effective males, queens or egg-laying females, and workers or ineffective females, in which the development of the reproductive organs is arrested or modified.   Distinct modes of behaviour are correlated with these structural differences.   When a swarm of bees leaves a hive it generally consists of the old queen-mother and a certain number of the workers which are her offspring. When they have found new quarters, or have been safely housed under domestication, the workers busy themselves in making the cells in which the queen may lay her eggs, and in which food may be stored.   In doing this the bees act in concert,

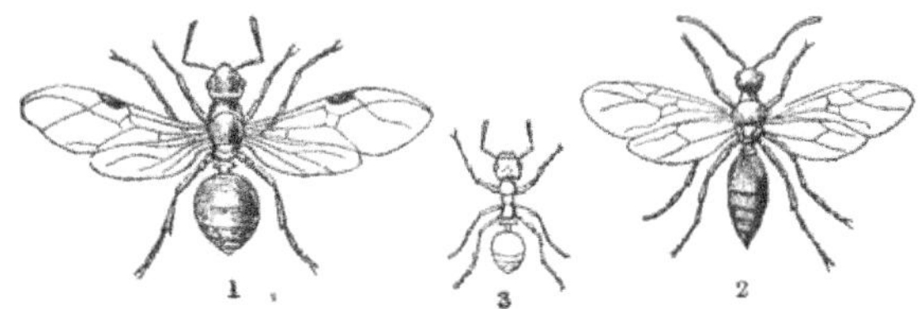

FIG. 24.—Wood ant.   1, Queen ; 2, male ; 3, worker (from Shipley).

and though the mathematical accuracy of the form and size of the cells has been much exaggerated, the comb which results is a very beautiful and well-adapted product of mutual co-operation in joint labour.   And though intelligence may, under special circumstances, modify the method of procedure there can be little doubt that comb-building is primarily due to inherited instinct.   The cells are not, however, all of the same size, those for the drones being somewhat larger than the cells in which the workers are reared, while much larger and differently shaped cells are prepared for the future queens. If instinctive therefore—as it seems to be in the main—the behaviour runs into different lines, the immediate causes of which, internal or external, we are not able accurately to assign.

The reproductive behaviour of egg-laying in the queen-mother is also instinctive.   It is believed that the drones are developed from eggs from which the queen bee withholds the fertilizing fluid, which she retains for months or years after the nuptial flight, stored in a special receptacle. And the size and shape of the drone-cell may supply the stimulus through which her behaviour in this respect is determined. But she lays similarly fertilized eggs in both the worker-cells and the queen-cells; and in these two cases the stimulating conditions must be different.

When the eggs have been laid, and the grubs hatched, the worker bees assume new duties—the feeding and tending of the young. They eat honey and pollen, which is partially digested, and supplied as pap to the grubs in such quantities that they seem bathed in it ; but after a short time a mixture of honey pollen and water is substituted for this pap. It is said that the drone larvæ are fed with pap for a longer period than the workers ; and the queen larva undoubtedly receives far more of this pap—or, perhaps, of a still richer nutritive product, sometimes spoken of as royal jelly—and, indeed, is supplied therewith throughout larval life. It is generally believed that this high feeding is the cause of queen-development, and that should the queen larvæ die ordinary worker larvæ are fed up, and produce queens nowise dissimilar to those developed in the royal cells. It is clear, if this be so, that the behaviour of the nurses decides the difference between the future queens and working bees—that is to say, the fertile and the sterile females. In any case, the feeding of the young by members of the same community is a fact to be specially noted. It is commonly said that the family is the germ from which the social community springs ; and it may be added that food-collection or food-administration in some form makes the difference between the family that coheres and the family that scatters.

When the larvæ have been fed, each after its kind, the workers seal up the cells with lids of pollen and wax ; the larvæ spin cocoons, pass into the pupa stage, and then change to perfect bees, which bite a way through the lid and take their place in the hive. These young bees now become the nurses, while the older bees go abroad to fetch honey and pollen to be stored away in some of the cells. But when a queen emerges, her first act is to go round to the other royal cells, tear them open, and sting to death the helpless occupants. Meanwhile the old queen may have led off the surplus population in a swarm, and the new queen reigns in her stead. Idle drones have also been emerging from their cells ; and when the young queen starts forth on her nuptial flight she is followed by the drones, mates with one of them, and returns a potential mother of thousands. So long as there is abundance of food the useless drones are tolerated ; but when there is scarcity they are ejected, and drone eggs, larvæ, and pupæ are said to be destroyed.

In the works of Huber and others, further marvels of hive-life, some well-authenticated, others more or less doubtful, are duly set forth. But enough has here been said to show that a social community of bees presents problems of animal behaviour which are sufficiently difficult of explanation. How far is the behaviour instinctive ? How far is it due to experience individually acquired ? Are we constrained to admit a rational factor ? If so, is it, like human reason, the result of generalization from experience of the relationships of phenomena ? Or are there features of insect psychology which differ from any of which we have firsthand knowledge ? These questions are more easily put than answered. As in the case of bird-migration, so too in that of the social life of bees, there is much that honesty forces us to confess our inability satisfactorily to explain.

So, too, is it in the social life of ants. Among these insects the males and perfect females bear wings, though these appendages may be subsequently shed. In some kinds, however, there are also wingless males or females capable of exercising the reproductive function. The workers are wingless, and are often of two or three kinds, differing in form and appearance, and in some cases playing different parts in the social economy. There is also, in some cases, a separate class of large-headed soldier ants ; so that differentiation of structure among the sterile females is carried further in ants than in bees. Their nests generally consist of an elaborate system of chambers and passages, either built with pine-needles, as in our common wood ant, or hollowed out in the earth or in wood, or sometimes built with a paper-like material, or formed of rolled leaves. It is said that a common ant in Eastern Asia (*Œcophylla smaragdina*) "forms shelters on the leaves of trees, by curling the edges of leaves and joining them together. . . . The perfect ant has no material with which to fasten together the edges it curls ; its larva, however, possesses glands that secrete a supply of material for it to form a cocoon with, and the ants utilize the larvæ to effect their purpose." * This has recently been confirmed by Mr. E. G. Green, Government entomologist, at the Botanic Gardens at Peradeniya, Ceylon. "He has seen ants actually holding larvæ in their mouths and utilizing them as spinning machines. To find out what would be done, some leaves were purposely separated by Mr. Green. The edges of the leaves were quickly drawn together by the ants, and, about an hour later, small white grubs were seen being passed backwards and forwards across the gaps made in the walls of the shelter. A continuous thread of silk proceeded from the mouth of the larva, and was used to repair the damage." † This is a remarkable act of apparently intelligent behaviour. But when we remember how much of the time of ants is occupied in carrying about their larvæ, it is hardly an act of which it can be affirmed that it could not arise as the result of chance experience.

In some cases two different genera are found in the same nest, with separate chambers and passages, as in the case of the robber-ant (*Solenopsis*) and the slave-ant (*Formica fusca*). The orifices by which the former enter are too small to allow of the entrance of the latter, "hence the robber obtains an easy living at the expense of the larger species," for "they make incursions into the nurseries, and carry off the larvæ as food."

In a few cases the foundation of a new colony has been carefully watched. Blockmann was successful in observing the formation of new nests by *Componotus ligniperdus* at

* Sharp, "Insects," part ii., p. 147.
† *Nature*, vol. lxii., p. 253 (July 12, 1900).

Heidelberg. " He found under stones, in the spring, many examples of females, either solitary or accompanied only by a few eggs, larvæ, or pupæ. Further, he was successful in getting isolated females to commence nesting in confinement, and observed that the ant that afterwards becomes the queen, at first carries out by herself all the duties of the nest. Beginning by making a small burrow, she lays some eggs, and when these hatch, feeds and tends the larvæ and pupæ : the first specimens of these latter that become perfect insects are workers of all sizes, and at once undertake the duties of tending the young and feeding the mother, who, being thus freed from the duties of nursing and of providing food while she is herself tended and fed, becomes a true queen-ant. Thus it seems established that, in the case of this species, the division of labour found in the complex community does not at first exist, but is correlative with increasing numbers of the society. Further observations as to the growth of one of these nascent communities, and the times and conditions under which the various forms of individuals composing a complete society first appear, would be of considerable interest." *

The queen does not, as in the case of the bee, deposit her eggs in separate cells where they are tended by nurses. The eggs, which are laid in the chambers of the nest, are subjected to much licking by the nurses ; the larvæ are, moreover, moved about from place to place, so as to be subjected to the requisite conditions of moisture and temperature. They are carefully cleaned, and after they have passed into the pupa stage the emerging insects are stripped of a delicate investing skin. And not only do the ants assiduously feed their young ; those who have gone forth and drunk their fill of sweet juices feed those who have remained behind. Forel took some specimens of *Componotus ligniperdus*, " and shut them up without food for several days, and thereafter supplied some of them with honey, stained with Prussian blue ; being very hungry, they fed so greedily on this that in a few hours their hind bodies

---

* The quotation is from " The Cambridge Natural History," vol. vi., " Insects," part ii., by David Sharp, F.R.S. ; see pp. 145, 146.

were distended to three times their previous size. He then took one of these gorged individuals, and placed it among those that had not been fed. The replete ant was at once explored by touches of the other ants and surrounded, and food was begged from it. It responded to the demands by feeding a small specimen from its mouth, and when this little one had received a good supply, it in turn communicated some thereof to other specimens ; while the original well-fed one also supplied others, and thus the food was speedily distributed. This habit of receiving and giving food is of the greatest importance in the life-history of ants." * It affords the basis or starting-point of the keeping of aphides, the making of slaves, the curious development of honey-pot ants, and in some cases the association with ants of other insects.

Some of these insects, of which there are many species belonging to several orders, are parasitic ; others appear to be hostile, and yet are able to maintain themselves in the nest ; others simply live side by side with the ants, which seem to be neither hostile nor friendly to them. In some of these cases the biological purpose of the association is unknown, while in others the ant serves as a model which the associated insect mimics. Thus in the nest of an Indian ant (*Sima rufa-nigra*) occur a small wasp and a spider which, to some extent in form and more markedly in coloration, mimic their hosts. "Wherever you find this species in any numbers," says Mr. Rothney,† " if you watch a few moments, you will see a mimicking spider, *Salticus*, running about among the ants, which it very closely resembles in appearance, much more so in life than in set specimens placed side by side ; I have seen numbers on the most friendly footing with the ants, though I have never seen them enter their burrows. . . . They are, I should say, the only friends the ant has, with the exception of a sand-wasp, a new species of *Rhinopsis* since described by Mr. Cameron, which also very closely mimics the ant, and which, on first

* Sharp, *op. cit.*, p. 147.

† G. A. J. Rothney, "Notes on Indian Ants," *Trans. Ent. Soc.*, 1889, p. 354.

observing among the workers, I took to be the male." But
there are some beetles which are not only tolerated, but fed
by the ants with which they live.   In the case of the genera
Atemeles and *Lomechusa*, which are always found in or near
ants' nests, the good offices are reciprocal, for the beetles
" have patches of yellow hairs, and these secrete some substance
with a flavour agreeable to the ants, which lick the beetles
from time to time.   On the other hand, the ants feed the
beetles ; this they do by regurgitating food, at the request of
the beetle, on to their lower lip, from which it is then taken
by the beetle.   The beetles in many of their movements
exactly resemble the ants, and their mode of requesting food,
by stroking the ants in certain ways, is quite ant-like.   So

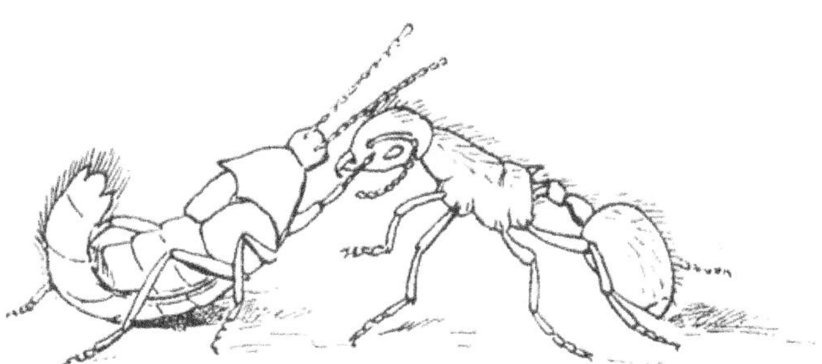

FIG. 25.—Beetle soliciting food from Ant (after Wasmann).

reciprocal is the friendship, that if an ant is in want of food
the beetle will in its turn disgorge for the benefit of its host.
The young of the beetles are reared in the nests by the ants,
who attend to them as carefully as they do to their own young.
The beetles are, however, fond of the ants' larvæ as food, and,
indeed, eat them to a very large extent, even when their own
young are receiving food from the ants.   Wasmann (to whom
we are indebted for most of our knowledge on this subject)
seems to be of opinion that the ants scarcely distinguish
between the beetle larvæ and their own young ; one unfortunate result for the beetle follows from this, viz. that in the
pupal state the treatment that is suitable for the ant larvæ

does not agree with the beetle larvæ. The ants are in the habit of digging up their own kind, and lifting them out and cleaning them during their metamorphosis : they do this also with the beetle larvæ, with fatal results ; so that only those that have the good fortune to be forgotten by the ants complete their development." *

Aphides, or plant-lice, yield to the solicitations of ants, which stroke them with their antennæ, by emitting a drop of sweet and viscid secretion, and it appears that the caress of the ant is the natural stimulus for the emission of the drop. Not only, however, do the ants go forth in search of aphides in their natural haunts, they bring them to the neighbourhood of the nest, and may even impound them by building a wall of earth round and over them. Huber stated that ants collected the eggs of the aphides and tended them in their nests, and the accuracy of the observation has been shown by Lord Avebury and others. "The aphid eggs are laid early in October, on the food plant of the insect. They are of no direct use to the ants, yet they are not left where they are laid, where they would be exposed to the severity of the weather and to innumerable dangers, but brought into the nests by the ants, and tended by them with the utmost care through the long winter months until the following March, when the young ones are brought out and again placed on the young shoots of the daisy." † Dr. McCook noticed that ants, returning from the trees on which aphides abounded, fed others near the nests, and he regarded this as a case of division of labour, the foragers obtaining food for the nurses which remained in or near the nest.

A further division of labour, carried to lengths which seem almost absurd, is found in the honey-pot ant of the United States and Mexico. The juice on which these ants feed is obtained from an oak-gall. Foragers go forth at night and return distended with the sweet fluid, and, having fed the

---

* Sharp, *op. cit.*, p. 226.

† Lord Avebury (Sir John Lubbock), quoted in Romanes' " Animal Intelligence," pp. 62, 63.

ordinary workers in the nest, apparently discharge the balance
of their store into living honey-pots, which remain in the nest
and preserve the food till it
may be required by the members of the community. Their
abdomens are enormously distended, they never leave the
nest, and they seem to form
a distinct caste, whose function
it is to passively accumulate
stores of reserve food for the
community. Curiously enough

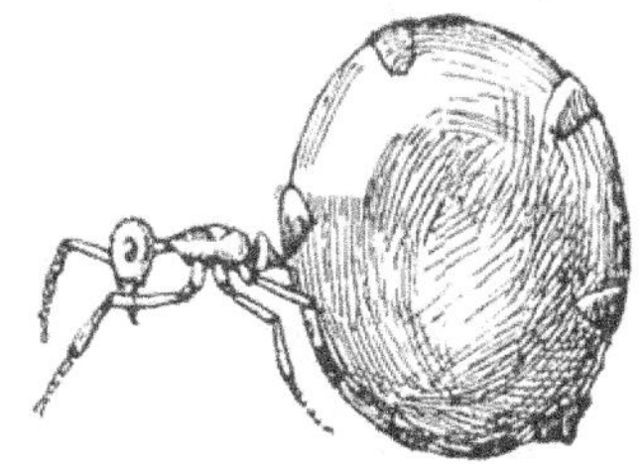

FIG. 26.—Honey-pot Ant.

the same peculiar social arrangement is found in different
genera living as far apart as Mexico, Australia, and South
Africa.

There is no doubt that in some cases the division of labour
is not restricted to the individuals of the same species, but that
other species are introduced into the nest to perform certain
functions—thus giving rise to the so-called slavery among
ants. This is carried to an extreme in the European species
Formica rufescens, the males and queens of which do no work,
while the sole function of the workers is to capture slaves of
the smaller species *Formica fusca*. In association with this
specialized mode of instinctive behaviour, " even their bodily
structure has undergone a change ; their mandibles have lost
their teeth, and have become mere nippers, deadly weapons
indeed, but useless except in war. They have lost the greater
part of their instincts : their art—that is, the power of building ;
their domestic habits—for they take no care of their own young,
all this being done by the slaves ; their industry—they take no
part in providing the daily supplies ; if the colony changes the
situation of its nest, the masters are all carried by the slaves to
the new one ; nay, they have even lost the habit of feeding. . . .
I have had a nest of this species under observation for a
long time, but never saw one of the masters feeding. I have
kept isolated specimens for weeks, by giving them a slave for
an hour or two a day to clean and feed them, and under these

circumstances they remained in perfect health, while, but for the slaves, they would have perished in two or three days." *

In this matter, we have in different species successive stages in the development of the instinctive behaviour which is thus carried so far in *Formica rufescens*. Our English ants, of the species *Formica sanguinea*, have fewer slaves and are less dependent on them ; they can feed and forage for themselves, and during migration carry their slaves—which are of the same species as in the other case—instead of being carried by them. In the nests of the common wood ant or horse ant (*Formica rufa*) there are occasionally a few slaves. Lord Avebury thinks it likely that they are developed from larvæ or pupæ, originally taken for food, which have by chance come to maturity in the nest of their captors.

But one more incident in the social life of ants can here be noticed—though many others could be given did space permit. The leaf-cutting ants of America form paths from their nests to suitable trees, from which to obtain the small coin-like leaf fragments, which they carry in the mandibles, and hence have gained the name of umbrella or parasol ants. These paths are sometimes underground ; and Mr. McCook measured one which ran at a depth of some 18 inches beneath the surface for 448 feet, and was then continued for another 185 feet to the tree which the ants were stripping. The whole path was in an almost perfect straight line from nest to tree. The leaf fragments are stored in large quantities in the nest, and it was long a matter of uncertainty for what purpose they were collected. The problem was solved by Alfred Möller, who found that the leaves, which are subdivided and masticated by a special set of workers within the nest, form the appropriate material in which the threads of a fungus ramify and flourish. This fungus is tended by the ants with great care, and is made to produce a specially modified form of growth, not found under other circumstances, in the form of white aggregations, termed by Möller "Kohlrabi clumps." These form the principal food of the ants ; and the spongy mass of earth and

* Lord Avebury (Sir John Lubbock), "Scientific Lectures," pp. 78, 79.

leaves is called the fungus garden. " If a nest be broken into
and the fungus garden scattered, the ants collect it as quickly as
possible, especially the younger parts, taking as much trouble
over it as over the larvæ. They also cover it up again as soon
as possible to protect it from the light." *

Again, it may be asked with regard to the social life of ants
as with respect to that of bees—How far is their complex
behaviour instinctive ? How far is it due to imitation ? What
part does intelligence play, and under what conditions of
acquisition ? Is reason, in the restricted sense of the word, a
factor in the development of the behaviour ? I cannot answer
these questions, and am of opinion that much detailed observation is yet needed before we can do much more than speculate
in the matter. Much indeed has been done, but yet more
remains for future investigation.

The conditions under which much of the behaviour is
carried out seem to indicate strong instinctive tendencies
which give an hereditary trend to the direction which the
social behaviour takes. Dr. Bethe,† indeed, goes so far as to
regard the behaviour as almost entirely instinctive, affording
little evidence of that modifiability of reactions which indicates intelligent guidance. He shows as the result of careful
experiment that the behaviour of ants to friends and enemies
are direct reactions to smell. Enemies washed with the excre-
tions of members of the nest are treated as friends, notwithstanding their different colour, size, and general appearance.
By scent, too, they follow the lead of others and retrace their
way to the nest ; this, he says, is not the result of a mental
process, but is the reaction of a complicated reflex mechanism.
As the outcome of careful observation, Dr. Bethe's conclusions
are of great value and interest. But he seems to go too far
in denying to ants any power of intelligent accommodation to
circumstances. If we admit intelligence, then the fact that

* *Nature*, vol. xlvii., p. 393 (Aug., 1893), where A. Möller's
investigations are described by J. C. Willis.

† A. Bethe, " Dürfen wir den Ameisen und Bienen psychische Qualitäten Zuschreiben," *Pflüger's Archiv.*, lxx., 1898.

the insects come forth in the midst of a community in full social activity would tend to the imitative or intelligent acquisition of like modes of procedure. It is difficult to distinguish the share taken by these two factors which may well cooperate. And if natural selection is exercising its influence through the elimination of those which do not fall into line in social behaviour, there would be ample opportunity for the survival of coincident variations.* If one may be allowed to speculate, it seems probable that the interaction of instinct and intelligence will be found with fuller knowledge to suffice for the explanation of the facts, without calling in the known but here improbable factor of rationality or any factors unknown elsewhere in psychology.

Some interesting observations of Lord Avebury's are sometimes quoted as evidence that ants are lacking in intelligence, but (if we accept the distinction already drawn †) they seem rather to show the lack of reason. " I placed food," he says, ‡ " in a porcelain cup, on a slip of glass surrounded by water, but accessible to the ants by a bridge, consisting of a strip of paper two-thirds of an inch long and one-third wide. Having then put an ant *(Formica nigra)* from one of my nests to this food, she began carrying it off, and by degrees a number of friends came to help her. When about twenty-five ants were so engaged, I moved the little paper bridge slightly, so as to leave a chasm just so wide that the ants could not reach across. They came to the edge and tried hard to get over, but it did not occur to them to push the paper bridge, though the distance was only about one-third of an inch, and they might easily have done so. After trying for about a quarter of an hour they gave up the attempt and returned home. This I repeated several times. Then, thinking that paper was a substance to which they were not accustomed, I tried the same with a bit of straw one inch long and one-eighth of an inch wide. The result was the same. Again, I placed particles of food close to and directly over the nest, but

* See p. 37.   † *Supra*, p. 138.
‡ " Scientific Lectures," pp. 80–82.

connected with it only by a passage several feet in length. Under these circumstances it would be obviously a saving of time and labour to drop the food on to the nest, or at any rate to spring down with it, so as to save one journey. But though I have frequently tried the experiment, my ants never adopted either of these courses. I arranged matters so that the glass on which the food was placed was only raised one-third of an inch above the nest. The ants tried to reach down, and the distance was so small that occasionally, if another ant passed underneath just as one was reaching down, the upper one could step on to its back, and so descend; but this only happened accidentally, and they did not think of throwing the particles down, nor, which surprised me very much, would they jump down themselves. I then placed a heap of mould close to the glass, but just so far that they could not reach across. It would have been quite easy for any ant, by moving a particle of earth for a quarter of an inch, to have made a bridge by which the food might have been reached, but this simple expedient did not occur to them."

Now, when we remember that the method of intelligence is to profit by chance experience, while the method of reason is, with foresight and intention, to adapt means to ends, we shall see that to move a straw even a quarter of an inch, or to make a bridge with particles of mould, would require rational and not merely intelligent powers. Chance experience would not supply the necessary data to be utilized by intelligence when repetition had established an association in the conscious situation. Granting that the ants were intelligent but not rational, they could not be expected to overcome the difficulties, simple as they seem to us, which Lord Avebury placed in their path. Had they been overcome the fact would be more difficult to explain than the use of a stone tool by the sand wasp, since this could more readily be hit upon by chance experience. And what these valuable experiments, of which kind more are needed, seem to show is, that the ant, probably the most intelligent of all insects, has no claim to be regarded as a rational being.

IV.—Animal Tradition

In that interaction between instinct and intelligence which, when further detailed work has sifted and purified our knowledge of the psychology of animal communities, may prove sufficient to account for the well-established facts, animal tradition will probably have to be recognized as of no little importance. When a newly emerging ant or bee, or a young bird or mammal is born into a community where certain modes of behaviour are already in full swing, an imitative tendency of the follow-my-leader type may lead it to fall in line with the traditional habits. It is said that young ants follow the older workers about the nest, and are " trained to a knowledge of domestic duties, especially in the case of larvæ." On the other hand, we have seen that, in certain observed cases, the queen ant is the solitary starting-point of a new community, and that the division of labour follows with the increasing numbers of the newly formed social group ; so that, in such cases, whatever part tradition may play in the later phases of social life, it cannot afford a sufficient account of the division of labour in the earlier history of the community. We need, however, fuller information concerning the continued life-history of such communities under natural conditions, and as to how far they remain self-contained without any incorporation of older members from adjoining nests. In the case of bees, where the old queen departs with a swarm, there may be greater continuity of tradition. But how far this is a necessary factor in social development is at present a matter of conjecture. In the herd of mammals and the flock of birds, and in all the family and social life in these classes of animals, the example of elders, without any imitation of the higher reflective type, can scarcely be without its influence on the behaviour of the young which, one would suppose, would tend to fall in with the ways which had become traditional in the species. Professor Wesley Mills tells us that a mongrel pup, whose psychical development he carefully watched, showed " extraordinarily rapid " progress when he was introduced to the society of other dogs, and was thus subjected to the influence of canine tradition.

How far this influence extends in animal communities—how far it is either a necessary or even an important contributory factor in the development of certain modes of behaviour—is at present in large degree a matter of speculation. And the only justification for speculation in science is that it may open our eyes to modes of influence the range and limits of whose effects may be submitted to the touchstone of careful observation, and, if possible, experiment. In this instance it is rather the indefiniteness of the evidence before us than its absence that stands in the way of any profitable discussion of the problem from the evidential point of view. And this indefiniteness is partly due to the fact that the need of observation is not realized, because this factor in animal behaviour has not been distinguished with sufficient clearness. It is worth while, therefore, to devote a short space to a consideration of the relation of this tradition to instinct and intelligence with a view to the focussing of observation on the facts by which it may be further elucidated.

In the first place, it is probable that, as in other modes of animal behaviour, traditional procedure is founded on an instinctive basis. This must be an imitative tendency of the broad follow-my-leader type indicated in the first section of this chapter. And this would afford wide instinctive foundations, which would owe their hereditary character to the fact that, under natural selection, those individuals in the community would survive which fell into line with the adaptive behaviour of their companions, while those which failed in this respect would be eliminated as more or less isolated outsiders, standing apart from the social life. In illustration we may take a hypothetical case, founded, however, upon observation. The Rev. S. J. Whitmee, a missionary in Samoa, believes that the tooth-billed pigeon of these islands (*Didunculus strigirostris*) " has probably been frightened when roosting, or during incubation, by attacks of cats, and has sought safety in the trees. Learning, from frequent repetition of the fright, that the ground is a dangerous place, it has acquired the habit of building, roosting, and feeding on the high trees ; and this habit is now operating for the preservation of this interesting bird, which a few years ago was almost extinct." * Now, in this case, the young birds which followed the lead of those who, under experience, had acquired the habit, would stand a better chance of survival than those who, failing to do so, were caught napping on the ground. In further illustration, we may take the case of two species of rats found by Mr. C. M. Woodford on one of the Solomon Islands. These two species are regarded by Mr. Oldfield Thomas as slightly altered descendants of one parent species, with adaptations due to the fact that, of this original species, some have adopted a terrestrial, others an arboreal life. Thus *Mus rex* lives in trees, has broad footpads, and a long rasp-like, probably semi-prehensile tail ; while *Mus imperator* lives on the ground, has smaller pads, and a short smooth tail. How far the different modes of behaviour in the two species may have been fostered by the influence of tradition we do not know ; but it is not improbable that such an influence would be a co-operating factor in the process of segregation, and that in the course of time each form has been adapted to its special environment through the elimination of those individuals which were not in harmony with the conditions of their life.

Such a case—admittedly hypothetical in the interpretation put upon the facts—may help us to see how the general instinctive follow-my-leader tendency might become specialized in certain essential lines of racial behaviour, and how, under natural selection, coincident variations in the line of traditional acts might become more and more definitely inherited as, at first, strong instinctive tendencies, and eventually more stereotyped modes of instinctive behaviour. This, indeed, may have been the mode of origin of some of the social instincts.

Reverting, however, to the stage where the general instinctive follow-my-leader tendency is only partly or incompletely

* *Proceedings of the Zoological Society*, 1874, p. 184.

specialized along particular lines of behaviour, we should have at this stage certain hereditary trends of action, dependent on stimuli afforded by the behaviour of others, but needing, for their guidance to finer issues and more adequate and highly perfected performance, the play of intelligence and the satisfaction of nascent social impulses. In the economy of the hive or the nest there are, no doubt, instinctive tendencies and predispositions; but there is also something more than organic heredity with its transmitted modes of behaviour analogous to the inherited form and structure of the body or its parts. Consciousness exerts a guiding influence. The insect is not independent of experience, but is capable of profiting by the teachings of that fertile mother of all intelligent behaviour. It is unnecessary, however, to insist on the fact that such insects are something more than instinctive automata, but are guided in their behaviour by the results of experience. Many careful observers lay stress upon this; if, indeed, they do not go further and claim for the social insect the higher rational faculty. "When we see," says Lord Avebury,* "an ant-hill tenanted by thousands of industrious inhabitants, excavating chambers, forming tunnels, making roads, guarding their home, gathering food, feeding the young, tending their domestic animals—each one fulfilling its duties industriously, and without confusion—it is difficult altogether to deny to them the gift of reason; and the preceding observations tend to confirm the opinion that their mental powers differ from those of man, not so much in kind as in degree."

If the term " reason " be here accepted in the broad sense, and not in the narrower sense before indicated, this passage will probably be endorsed by the majority of those who have paid any attention to the subject; save that those who regard "reason," in the more restricted acceptation of the term, as outside any scheme of evolution, may claim that this does constitute a difference in kind and not merely in degree. In any case the passage expresses the conviction of a close and

* "Scientific Lectures," 2nd edit., p. 140.

singularly unprejudiced observer, that the doings of ants involve conscious guidance in the light of experience individually acquired.

And yet the behaviour of different species of ants, each after its kind, is remarkably constant—so constant that, to use the words of Dr. Peckham in another connection, it is characteristic of the species, and would be an important part of any definition of the insect based upon its habits. And some part of this constancy may be due to tradition, though much of it may result from strong instinctive tendencies which intelligence guides to similar ends, because the conditions are similar in successive generations of social insects.

From the point of view of observation, however, it is particularly difficult to distinguish the part played by tradition as a psychological influence from that played by what we have above described as instinctive imitation. In our study of other modes of instinctive behaviour we can isolate an individual, or group of young individuals, and observe how far certain acts are performed prior to any experience. Thus chicks behave in certain instinctive ways under conditions which preclude their learning from the hen or other older birds—so that tradition cannot be operative. But where social behaviour is concerned, such methods of observation are necessarily excluded—save in such cases as that of the incipient community of ants. For if certain instinctive acts require for their due performance the stimulus of the like performance in others, what is this but a form of instinctive tradition ; and how are we to distinguish it from intelligent tradition, where a psychological factor has freer play and exercises guidance over the performance ? In the present state of our knowledge we can do no more than suggest, as not improbable, that tradition passes through three phases : the first in which it is instinctive ; the second in which it becomes intelligent through the satisfaction which the due performance of traditional acts arouses in consciousness ; and the third in which, at any rate in man, it takes on a rational form, and is made to accord with an ideal scheme, the product of conceptual thought and of reflection on data which have been generalized and considered in their due relationships to the scheme which takes definite form in the mind. Whether in the social communities of insects or those of beavers, among mammals, or rooks among birds, tradition has begun to pass into the third or rational stage, we do not know. It may be so, but probably the development along these lines has not been carried far. Presumably in the ant, rook, and beaver anything like an ideal scheme of thought based on reflection, if it exist, is as yet exceedingly indefinite.

But even supposing that no animal has yet risen beyond the second or intelligent stage, it is none the less important to realize that we have here, in animal life, the foundations on which may be raised what may, perhaps, be regarded as one of the characteristic features of human progress. This characteristic is the transference of evolution from the organism to the environment handed on from generation to generation. Thus man, "availing himself of tradition, is able to seize upon the acquirements of his ancestors at the point where they left them." * Thus "he has slowly accumulated and organized the experience which is almost wholly lost with the cessation of individual life in other animals." † But he is able to do so through the extension, refining, and fixing of that instinctive and intelligent tradition which begins to take form in animal communities.

V.—THE EVOLUTION OF SOCIAL BEHAVIOUR

"Animals of many kinds," said Darwin,‡ "are social; every one must have noticed how miserable horses, dogs, sheep, etc., are when separated from their companions. The most common mutual service in the higher animals is to warn one another of danger. Every sportsman knows how difficult

* Weismann, "Essays," vol. ii., p. 50.

† Huxley, "Collected Essays," vol. vii., p. 155.

‡ "The Descent of Man," vol. i. p. 853, 2nd Ed., 1888. The quotations from Darwin in this paragraph and that which follows are somewhat condensed by a few omissions.

it is to approach animals in a herd or troop. Wild horses and
cattle do not, I believe, make any danger-signal; but the
attitude of any one of them who first discovers an enemy
warns the others. Rabbits stamp loudly on the ground with
their hind feet as a signal; sheep and chamois do the same
with their fore feet, uttering likewise a whistle. Many birds
and some mammals post sentinels. The leader of a troop of
monkeys acts as such, and utters cries expressive both of
danger and of safety. Social animals perform many little
services for each other : horses nibble, and cows lick each
other ; monkeys search each other for external parasites, and
are said to remove thorns and burrs. Social animals mutually
defend each other. Bull bisons in North America, when there
is danger drives the cows and calves into the middle of the
herd, whilst they defend the outside. Among baboons the old
males come forward to the attack. Wolves hunt in packs ;
and pelicans fish in concert.

"It has often been assumed," continues Darwin, "that
animals were in the first place rendered social, and that they
feel as a consequence uncomfortable when separated from each
other, and comfortable whilst together ; but it is a more probable view that these sensations were first developed, in order
that those animals which would profit by living in society
should be induced to live together, in the same manner as the
sense of hunger and the pleasure of eating were, no doubt, first
acquired in order to induce animals to eat. The feeling of
pleasure from society is probably an extension of the parental
and filial affections, since the social instinct seems to be
developed by the young remaining long with their parents ;
and this extension may be attributed in part to habit, but
chiefly to natural selection. With those animals which were
benefited by living in close association, the individuals which
took the greatest pleasure in society would best escape various
dangers ; while those which cared least for their comrades, and
lived solitary, would perish in greater numbers. In however
complex a manner the feeling of sympathy may have originated,
as it is one of high importance to all those animals which aid

and defend one another, it will have been increased through natural selection ; for those communities which included the greatest number of the most sympathetic members would flourish best, and rear the greatest number of offspring."

It is impossible to improve upon this pithy description of the salient facts, and terse explanation in terms of the hypothesis of natural selection. It may, perhaps, be urged that, on this hypothesis, the origin of the social state, through a biological association of individuals, probably neither preceded nor followed the development of a psychical bond arising from the sense of satisfaction and comfort afforded by social life, but that both originated *pari passu*. If the linkage was primarily instinctive, its intelligent continuance could only be effected through the pleasure social behaviour carried with it, and the discomfort of separation from the community. No instinctive acts would be persistently repeated, under the guidance of individual experience, if that experience proved bitter and not sweet. An animal with thwarted instincts is one with unsatisfied impulses ; its biological and its psychological tendencies are alike unfulfilled. What Darwin saw and wished to enforce, however, was that the psychical link of conscious satisfaction was a necessary prerequisite of the continuance and further evolution of sociability ; and that without the integrating bonds of sympathy any advance of social development was impossible.

In two able and interesting articles in the *Nineteenth Century* review,* on "Mutual Aid among Animals," Prince Kropotkine gives a useful and sufficiently detailed summary of the chief facts concerning the social relationships which have been observed in the animal kingdom—including, perhaps, some rather apocryphal instances,—and combats Huxley's statement † that, " beyond the limited and temporary relations of the family, the Hobbesian war of each against all is the normal state of existence " among animals and primitive men. " Life

* Vol. xxviii., Sept. and Nov., 1890, pp. 337-354, 699-719.

† *Nineteenth Century*, Feb., 1888, p. 165. " Collected Essays," vol. ix., p. 204.

in societies," says Prince Kropotkine, " is no exception in the animal world. It is the rule, the law of nature, and it reaches its fullest development with the higher vertebrates. Those species which live solitary, or in small families only, are relatively few, and their numbers are limited." * " Life in societies enables the feeblest insects, the feeblest birds, and the feeblest mammals to resist, or to protect themselves from, the most terrible birds and beasts of prey ; it permits longevity ; it enables the species to rear its progeny with the least waste of energy, and to maintain its numbers, albeit with a very slow birth-rate ; it enables the gregarious animals to migrate in search of new abodes. Therefore, while fully admitting that force, swiftness, protective colours, cunningness, and endurance to hunger and cold, which are mentioned by Darwin and Wallace, are so many qualities making the individual or the species the fittest under certain circumstances, we maintain that under *any* circumstances sociability is the greatest advantage in the struggle for life. . . . The fittest are thus the most sociable animals, and sociability appears as the chief factor in evolution, both directly, by securing the well-being of the species while diminishing the waste of energy, and indirectly by favouring the growth of intelligence." † And summarizing his argument, Prince Kropotkine says,‡ " We have seen how few are the animal species which live an isolated life, and how numberless are those which live in societies, either for mutual defence, or for hunting and storing up food, or for rearing their offspring, or simply for enjoying life in common. We have also seen that, though a good deal of warfare goes on between different species, or even different tribes of the same species, peace and mutual support are the rule within the tribe, or the species ; and that those species which best know how to combine, and to avoid competition, have the best chances of survival and of further

* *Op. cit.*, pp. 709, 710.
† Page 711.
‡ "Mutual Aid among Savages." *Nineteenth Century*, vol. xxix., April, 1891, p. 538.

progressive development. They prosper, while unsociable species decay."

Prince Kropotkine seems, however, to push his argument too far. The assertion that the fittest are the most sociable animals, that sociability appears as the chief factor in evolution, and that unsociable species decay, is not likely to be accepted without qualification by zoologists. What grounds have we for saying that the solitary wasps are less fit than the social wasps ? Each has a fitness according to its kind. Can it be maintained that the unsocial tiger is less fit than the social jackal ? And can it be said that tigers, which are reported absolutely to swarm in Java and Sumatra, exemplify the decay of an unsociable species ? Is it seriously contended that the hawk, which may be successfully mobbed by a number of wagtails, is less fit than his more social assailants ? And are the unsocial raptorial birds decaying species ? Such questions might be asked by the score. And the answer in every case is that the social and unsocial alike are fitted to their several states of life. In fact, it might be contended, with every whit as much if not more cogency, that sociability is nature's device for enabling the weaker, and hence in themselves the less fit, to resist the attacks and encroachments of the stronger and individually fitter. Discussing the possibilities of human ancestry, Darwin said : * " In regard to bodily size or strength, we do not know whether man is descended from some comparatively small species like the chimpanzee, or from one as powerful as the gorilla. We should, however, bear in mind that an animal possessing great size, strength, and ferocity, which, like the gorilla, could defend itself from all enemies, would not perhaps have become social ; and this would most effectually have checked the acquirement of the higher mental qualities, such as sympathy and the love of his fellows. Hence it might have been an immense advantage to man to have sprung from some comparatively weak creature."

Zoologists, again, will hardly accept without question Prince Kropotkine's assertion that " life in societies is no

* " Descent of Man," vol. i., p. 96.

exception in the animal world, but is the rule, the law of nature." Many will contend, on the other hand, that life in societies with anything like division of labour, or with mutual aid (and this seldom carried far), is, taking the animal kingdom as a whole, of comparatively rare occurrence, though none the less noteworthy where it exists. And, in any case, it seems somewhat extravagant to say that sociability is the chief factor in evolution. No doubt it might be plausibly urged that human society is, from man's point of view, the highest product of evolution ; that in attaining to this end sociability has been the leading factor ; and that obviously the leading factor in the evolution of the highest product may properly be called the chief factor in evolution. But Prince Kropotkine apparently means that sociability is the chief factor, not only in this evolution, but in all organic, or, at least, all animal evolution. In this he will receive the support of but few zoologists. By some extravagance of statement he has weakened his own case, which is otherwise not lacking in points of weakness. The legitimate inferences from animal behaviour are, that co-operation is in some cases a factor in the evolution of a successful species, that in human progress it has been an important factor giving strength to a creature weak in tooth and claw, and that this factor has co-existed, and still coexists, with that of competition, in the absence of which the race would be dragged down to lower levels of efficiency by the incubus of weaklings.

To Professor Alfred Espinas * we owe the best and fullest discussion of the social life of animals, and to his work the reader may be referred for a careful and, for the most part, unstrained and unbiassed consideration of the phenomena. In common with others who have devoted serious attention to the subject, he sees in the family the starting-point of the higher and more comprehensive social group, or "peuplade." Prince Kropotkine seems, indeed, to combat this view ; but the divergence of opinion is more apparent than real. He

* "Des Sociétés Animales : Étude de Psychologie Comparée" (Paris, 1877).

tells us * that anthropology "has established beyond any doubt that mankind did *not* begin its life in the shape of small isolated families. Far from being a primitive form of organization, the family is a very late product of human evolution. . . . Societies, bands, or tribes—not families—were thus the primitive form of organization of mankind and its earliest ancestors." And in support of his views he adduces the sexual communism which is said to be found in the lowest savages, and briefly traces the development of monogamy and the genesis of the family ideal as we conceive it. It may at once be admitted that in all probability mankind did not have its origin in small isolated families. If we do not admit this we must accept the alternative hypothesis, that man was developed from an unsocial ancestor. For though the biological family is the starting-point of the community, it does not of course follow that wherever there is so much coherence between parents and offspring as to form a temporary family group, a social community must in due course arise. In such unsocial carnivora as the tiger, the temporary linkage of family life is strong while it lasts. But though mankind presumably originated in a prehuman race that had already reached some degree of social coherence, there remains behind the question—what was the origin of this social group? And to this question, Prince Kropotkine, in common with Darwin and Espinas, would probably answer without hesitation, that the primeval germ of the social community lay in the prolonged coherence of the group of parents and offspring. In the unsocial animals the family separates and disintegrates before the offspring mate. But if the family continue to cohere, the mating of offspring will give rise to the continuity of coherence found in the herd, or troop, or tribe. For new family groups will be constantly arising before the old family groups have ceased to be associated. Thus would be afforded more opportunity for tradition than among the unsocial animals.

* "Mutual Aid among Savages," *Nineteenth Century*, April, 1891, pp. 539, 540.

How, then, can it be said that, "far from being a primitive form of organization, the family is a very late product of human evolution"?   By using the word "family" in a sense somewhat different—nay, widely different—from that in which it is employed in a biological discussion.   In the latter usage sexual communism is not excluded ; A., B., and C., D. may have offspring this season ; A., D., and C., B. next season.   In each season there are family groups with interchange of partners.   This does not, however, conform with our conception of the family as realized under civilization.   Herein, in fact, lies the essential difference between the human and the animal family.   The one is a realized ideal ; the other is merely a natural occurrence. Even in the case of monogamous animals, mating for life is probably not conduct in conformity with an ideal, but is due to the fact that instinctive tendencies have taken this line of direction.   On the other hand, in monogamous communities of mankind, there is, unfortunately, evidence that in some cases the ideal is not strong enough to prevent presumably ancestral tendencies in the direction of communism.

The basis of human social conduct is unquestionably to be traced in the social behaviour of animals, in inherited tendencies to co-operation and mutual help, in the bonds of sympathy arising through the satisfaction of impulses towards such behaviour, and perhaps, to some extent, in the influence of tradition.   It is not, however, until this tradition is rendered, through descriptive communication, more continuous and more effective ; it is not until an ideal of mutual aid, and social conduct generally, takes form and is rendered common to the tribe ; it is not until the more or less realized conceptions of one generation are handed on to become the environment under which the succeeding generations are nurtured ; it is not indeed until man consciously and reflectively aims at the bettering of his environment in accordance with standards rationally conceived and deliberately carried into execution ; that a new *régime* of civilized progress, elsewhere unknown in nature, takes definite form.   Under this *régime*, the elimination of failures through natural selection, though it may not be entirely superseded, plays a subordinate part ; alongside the organic continuity which is due to physical heredity, there runs a continuity of tradition through social inheritance.

Human civilization is an embodiment of reason, a product of reflection, a realization of ideals conceived by the leaders of mankind. All this forms the environment of each one of us. And it is this environment which is undergoing progressive evolution and playing on the rational faculties of those which are submitted to its moulding influence. There is no sufficient evidence of anything of the kind in the social communities of animals. This, of course, must be accepted merely as an expression of opinion. But on the hypothesis that animals are rational beings, capable of reflection, it is difficult to understand why they should remain at so low a level of social achievement. The absence of powers of descriptive intercommunication is often assigned as the cause of their comparatively unprogressive condition ; but it may be regarded as the sign, rather than the cause, of their lack of reason in the more restricted sense of the term. We cannot, however, enter into the much-disputed question whether reason is the product of language, or language the outcome of reason. Perhaps the safest position is to assume that rationality and true speech are in large measure different aspects of one evolutionary movement—speech arising out of such preceding modes of communication as were considered in the second section of this chapter ; reason developing out of intelligence which supplies its necessary data. It is sometimes said that, notwithstanding their powers of speech, savages in their social relations show little advance on animal communities. But surely such statements must be made in forgetfulness of the fact that savage customs almost invariably indicate the presence and sway of ideals which puzzle us from their quaintness, and from the fact that they seem contrary to the dictates of intelligence and due to motives and conceptions the nature and force of which we find it difficult to estimate. The passage from intelligent social behaviour to the rationality which has assumed such strange aspects among existing savages took place somewhere at some time in the past ;

and the stages of its evolution are hidden from our view. All we can say is, that it is possible to trace in animal behaviour some of the instinctive tendencies and intelligent modes of accommodation to social circumstances, together with the germs of imitation, intercommunication, and tradition, and the establishment of bonds of sympathy, without which the subsequent stages of evolution would be inconceivable.

CHAPTER VI

THE FEELINGS AND EMOTIONS

I.—IMPULSE, INTEREST, AND EMOTION

ANY discussion of animal behaviour must deal largely with
what is termed the conative aspect of consciousness. "The
states designated by such words as *craving*, *longing*, *yearning*,
endeavour, *effort*, *desire*, *wish*, and *will*," says Dr. Stout, in his
admirable "Manual of Psychology," * "have one characteristic
in common. In all of these there is an inherent tendency to
pass beyond themselves and become something different. This
tendency is not only a fact, but an experience; and the peculiar
mode of being conscious, which constitutes the experience,
is called *conation*." Closely associated with this conation is
impulse, which Dr. Stout defines as "any conative tendency,
so far as it operates by its own isolated intensity, apart from
its relation to a general system of motives. Action on impulse
is thus contrasted with action which results from reflection and
deliberation." † In the interpretation we have advocated,
animals are essentially creatures of impulse, and not to any
large extent, if at all, reflective agents. And their impulses
may be associated either with their inherited and congenital
behaviour, or with that which is due to acquired experience.
In other words, their impulses may be divided broadly into
two classes, the one instinctive, the other acquired.

Dr. Stout says that conation is not only a fact, but an experience. Now, first as to the fact. It seems to be the

* Page 63.
† *Op. cit.*, p. 267; *cf. supra*, p. 138.

correlative in consciousness of the behaviour of the nervous system under stimulation. Let us take some simple case such as that, for example, of a hungry chick pecking at a grain of corn. This is explained from the physiological point of view by saying that internal and external stimulation—internal from the digestive organs and system in need of food, external through the eye—gives rise to a state of unstable equilibrium in the nerve-centres ; and that, when the instability reaches a certain value, the nervous system discharges into the motor organs, the chick pecks, and stable equilibrium is restored. The tendency to discharge in some way under stimulation is an essential characteristic of a nervous system. It is one of the facts of physiological science. So, too, the conative tendency is one of the facts of psychological science ; it is a change in the situation introduced by the effects of the physiological discharge.

Let us here parenthetically notice that the physiological tendency in the nervous system is an evolved complication and a specialized development of one of the fundamental properties of protoplasm—that which is often spoken of as irritability. One of the characteristics of all living matter is its "explosive" instability. So that at the very threshold of organic behaviour we have the analogue of that which, in its developed form, becomes the tendency of the nervous system to discharge as the result of stimulation. The conative tendency of the psychologist, therefore, has its roots deep down in the elemental germs of all organic life.

And this conative tendency is, says Dr. Stout, not only a fact but an experience. Let us return to our hungry chick that pecks at a grain of corn. And let us grant that as the result of stimulation there arise states of consciousness which we describe as a feeling of hunger and the sight of the grain of corn. The nervous system now discharges ; and there are introduced into the situation a further group of data, the motor consciousness of the actual behaviour, sensory data from the results of the act, the seizing of the grain and so forth. The situation has unquestionably changed. But is there any specific consciousness of the conative tendency as such ?   Is there any "peculiar mode of being conscious which constitutes the experience which is called conation"?   It is difficult to say.   Hence we find differences of opinion among psychologists as to what, from the psychological point of view, the impulse actually is.   Is it simply the conscious situation prior to the response?   Is it a feeling of the change from the initial to the succeeding phase?   Or are new data introduced apart from those afforded by stimulation on the one hand and response on the other?   We will not attempt to decide.   Without determining its exact nature we may rest content with the very general statement that impulse is a concomitant of a change in the conscious situation.

There is, however, a use of the term concerning which it seems necessarily to enter a word of protest.   Impulse is by some regarded as the underlying cause of the conative tendency. Now science, as such, has nothing whatever to do with underlying causes.   If, as a matter of observation and inference, we have reason to believe that there is such a tendency, science simply accepts the fact, and endeavours to formulate the conditions under which it arises, and to trace its observed or inferred antecedents.   No doubt many of us find it difficult or impossible to rest content with the strictly scientific position, that of unquestioning acceptance of the facts of nature as we find them given in experience.   We say : Here is an observed tendency the conditions and antecedents of which are described by science.   But what causes the tendency ; what is the impelling force ?   Now to such questions science can give no answer.   Science deals with phenomena, and tries to tell us all about their conditions and their antecedents.   But whenever Science is asked : " What is the underlying cause of the phenomena,—that which calls them into being ? " Science should always give one answer and one only : " Frankly, I do not know ; that lies outside my province ; ask my sister Metaphysics."   Science ought to have nothing whatever to do with force as the underlying cause of anything in this universe of phenomena.   And impulse, as the impelling force which calls a conative tendency into being, is a metaphysical, not a scientific conception.

We need not further discuss the psychological nature of impulse. Indeed, the little that has been said would not have been necessary to our inquiry were it not that we frequently have occasion to speak of animals as "creatures of impulse," and to refer to their behaviour as due to impulse. What do we mean by such expressions ? If we regard conative tendency as a fact (whatever may be said for or against its being also a specific experience), and if this fact is the tendency of the conscious situation to develope in certain definite ways, then we may define *impulse* with sufficient clearness by saying, with Dr. Stout, that it is characterized by being unreflective. Conative tendency thus comprises two categories—impulse and volition ; the one unreflective, the other involving deliberation.

Before passing on to consider how impulse is partly determined by the feeling-tone and the emotional attributes of the conscious situation, we may first draw attention to the important way in which the results of conative tendency afford the data through which consciousness attains its unity in the midst of diversity of experience.

We said that the impulses might be divided broadly into two classes—the one instinctive, the other acquired. Now, from the point of view suggested by a study of behaviour, if not also, as I am disposed to think, from the more general standpoint of a genetic study of mental development, it is convenient to start with the instinctive act and the conscious situation it implies. We have here a piece of experience which, if we may so phrase it, hangs together ; in which experience of things in the environment is included in the same elemental synthesis with that of bodily acts in organic relation to these things. It is closely linked, on the one hand, with a foregoing act of attention, itself of the instinctive type ; closely linked, on the other hand, with the results of behaviour through which the environing things call forth a new conscious situation and evoke a further response. Thus not only does the experience of an instinctive act hang together, but a series of such acts do so likewise. And coalescent association not only links and groups the elements within the situation called forth by the single act, but comprises also the elements of the developing situation afforded by the whole series. We see this in the young chick, where, as the result of experience, attention is emphasized where the material is palatable, and lapses where it is nauseous—such nauseous substances being soon ignored. Furthermore many environing things appeal in different ways to the same limited number of sense organs, while the same motor organs respond in different ways in successive modes of instinctive behaviour. The same brain forms the physical basis of varied situations overlapping in many ways, and receives afferent messages from the same body. Hence, in its organic unity it affords the conditions for an underlying stratum of mental unity, amid all the diversities of experience ; while the multiplicity of messages on the one hand from external things, and on the other hand from internal happen-ings, lays the foundations of a differentiation between the external world and the self—a differentiation long to remain implicit, and only to be rendered explicit on a far higher level of mental development. For at this early stage, and perhaps throughout animal life, " there is no single continuous self con-trasted with a single continuous world. Self, as a whole, uniting present, past, and future phases, and the world as a single coherent system of things and processes, are ideal con-structions, built up gradually in the course of human develop-ment. The ideal construction of self and the world is com-paratively rudimentary in the lower races of mankind, and it never can be complete. On the purely perceptual plane [with which we are now dealing] it has not even begun." * But though the ideal constructions of self and the world have not, as Dr. Stout says, at this stage, even begun, yet, as the same author observes,† " animals distinguish in the environment, and treat as a separate thing, whatever portion of matter appeals to their peculiar instincts, and affords occasion for their charac-teristic modes of activity." And this differentiation of specially

* G. F. Stout, " Manual of Psychology," p. 268.   † *Ibid.*, p. 318.

interesting things from each other, and from their relatively uninteresting surroundings, must be accompanied by some differentiation of these things from themselves as affected by them and reacting to them. So that here, as we have seen to be the case in other matters, what is commonly called the perceptual life of animals affords the rough-hewn materials from which ideal constructions may be elaborated by rational beings.

We cannot here attempt to do more than barely indicate the manner in which the perceptual process in animals may acquire unity and diversity—unity through the functioning of the same brain and body, diversity from the different modes of functioning and the differential effects of diverse modes of stimulation. The interesting point for us in our special inquiry is that it is through behaviour that all this is brought about.

As we interpret the facts, the restless activity of the young is primarily a biological fact, and is to be dealt with as an organic problem—a complication of the fundamental irritability of protoplasm. But it is also an essential condition to the acquisition of conscious experience ; and the more there is of it in varied modes the wider is the range of the data afforded to consciousness. Behaviour is thus the goal of organic heredity, and the starting-point of conscious accommodation and adjustment ; it is the biological end of variation, and affords the means to intelligent modification.

So much for some of the results of conative tendency. Not only does it secure adaptation or adjustment to the environment, but it affords the conditions of mental development by which further accommodation is rendered possible. But, in addition to the attainment of biological ends, in addition to the furtherance of survival in the struggle for existence, mental development has another aspect. All sensory data, whether from the special senses, from the motor processes concerned in responsive behaviour, or from other sources, may, and perhaps always do, carry with them some amount of what is termed *feeling-tone*, giving rise to a net result in consciousness

which we call pleasure or the reverse. Pleasure or satisfaction—however we name that which, though vague and
indeterminate in outline, is a very real attribute of the conscious situation—affords its sanction to certain modes of
conation, and may thus be regarded as the psychological end
of their continuance or their repetition.   It is partly, no doubt,
a direct adjunct of sight, hearing, taste, and so forth, and of
smooth and easy movements of the body and limbs; but it is
partly due to a great body of stimulation coming from many
parts of the organic system.   The blood-vessels are dilated
or contracted, the heart's action increased or diminished;
respiration is deepened or the reverse, and its rhythm may
be altered; glands are thrown into a state of activity;
the tone of the muscles is affected, and there may be either
incipient contraction or relaxation.   These are primarily
organic effects; but they influence the conscious situation, and
are themselves suffused with feeling-tone.   For, from all the
parts so affected, messages are carried in to the brain, and such
afferent messages afford data to consciousness.   It may be
that the experience of the conative tendency, for which
Dr. Stout and others claim a distinctive place in consciousness,
is largely due to afferent messages from the motor organs
incipiently innervated in preparation for the behaviour which
follows.   In any case these probably form very important
elements in the conscious situation antecedent to the actual
response.   In what we may term motor attention—the state
well exemplified by a cat in the strained pause which precedes
the spring on to the prey, or in ourselves when we poise
before a dive or hold the billiard cue in preparation for a
delicate stroke—this incipient innervation, felt through afferent
messages from the parts thus braced for action, enters with
much distinctness into the conscious situation.   In sensory
attention, on the other hand, reflex acts have actually taken
place, having for their end and purpose the focussing of the
sense organs on the object which stimulates them, so that in
this way further and more effective stimulation may be received.
But, as the sense organ is steadily held to the focus, and made

R

effectually to cover the stimulating thing, the motor apparatus concerned is kept on the strain, and is all the while contributing data to the conscious situation.

In primary genesis attention, both motor and sensory, is unquestionably organic and reflex in its nature. It is a product, and an invaluable product, of biological evolution. Without this as a basis, the higher forms of attention under conscious guidance would be impossible. For all these higher forms are modifications and complications of what is given in organic heritage. Here, as elsewhere throughout the whole range of behaviour, consciousness only guides to finer issues what is presented to it in rough outline, or in isolated fragments, as the outcome of biological evolution. But the organic responses afford the data which consciousness uses that it may mould and fashion the behaviour so as to reach higher and more complex modes of adjustment.

Lest a familiar form of words should give rise to misapprehension, it may here be stated that, when we say that consciousness moulds and fashions behaviour, we do not intend to imply that consciousness is an underlying cause. We are not using the term consciousness in a metaphysical sense. We mean that consciousness is the expression of certain conditions under which behaviour is guided. Instead of saying, therefore, that consciousness utilizes certain sensory data, it would be more correct to say that it *is* the sum-total of these data which are the psychical expression of certain brain conditions under which behaviour, as a matter of fact, takes a given set or direction. We use the word consciousness, then, not in its metaphysical sense of an underlying cause or force, but in its scientific sense, as the concomitant of certain antecedent conditions. Our common modes of speech lend themselves with misleading facility to metaphysical assumptions, all the more insidious since they are not consciously acknowledged as such. And not only what we comprise under the broader group-name "consciousness," but what we include under narrower group-names, such as "impulse," "volition," "instinct," "intelligence," "reason," and the like, often do duty as underlying causes of the phenomena, which, from the scientific point of view, they do no more than name.

We often say, for example, that *interest* guides behaviour in this direction or in that. But such interest must not be regarded as an impelling force ; it is an attribute of the conscious situation, more or less suffused with feeling-tone. It is not easy to define ; but it seems to take on its distinctive character when re-presentative elements contribute what Dr. Stout * terms "meaning" to the conscious situation. The meaning in the early stages of mental development is, however, merely perceptual, and not that which comes much later—that which is implied in the phrase "rational significance." In the chick which has tasted a cinnabar caterpillar the situation evoked by the sight of this larva has meaning in virtue of the actual experience. But, in this case, the meaning is not conducive to continued interest, since it checks, rather than stimulates, behaviour. At first, indeed, there may be the repellent interest of aversion. But this passes by, and the larvæ are soon ignored. Small worms also acquire meaning, and here the interest is attractive, and is stimulated afresh each time the meaning is reinforced by repetition of the act of seizing and swallowing.

We have seen that it is through behaviour that things become differentiated from their surroundings, and acquire relative independence in experience. It is through behaviour that what we have termed conscious situations develop. The thing is the centre or nucleus of a developing situation—that which starts the behaviour, and towards which the behaviour is directed ; or, since the behaviour may be that of avoidance or escape, we should, perhaps, rather say, it is that to which the behaviour has reference. Now, if interest is the feeling-tone attaching to the whole attentive situation, and if the nucleus of the situation is the thing, it naturally follows that the thing becomes the centre of interest. The mouse is a centre of absorbing interest to the cat, her eggs to the mother-bird, his mate to the sparrow in the spring. Companions are centres

* "Manual of Psychology," p. 84.

of abiding interest to social animals, because they are also the centres of social behaviour and the conscious situations arising thereout ; because they evoke in special ways the attentive situation.

The differentiated thing being thus a centre of interest, a relatively fixed nucleus in a changing conscious situation, the development of which is due to behaviour, there can be no question that, among social animals, the companion becomes a peculiar and specialized centre. Around him develops a particular type of behaviour. Towards him the reactions are of a quite distinctive kind. Mother and offspring, mate and mate, are reciprocal centres of interest. To the offspring the parent is a common centre of interest. As they grow up together, what is of interest to one is likewise of interest at the same time to others. Imitation begets similarity of conscious situations. In many ways such community of interest is fostered ; and through this community of interest the conscious situations acquire their distinctively social character. Not only is the companion, as the nucleus of a situation, a thing which reacts in altogether special ways, so that it becomes differentiated from other things as something the meaning of which, and the interest in which, are *sui generis* and unique in type ; but it enters into other situations in ways that are also peculiar and characteristic. A worm is thrown to a couple of chicks, and is to each a centre of interest—the nucleus of a situation involving appropriate modes of behaviour. But into this situation there enters for each of them, in a quite peculiar and distinctive way, the action and behaviour of the other chick. The situation is complicated by the introduction of a second centre of interest, and the behaviour has reference to both centres. Instead of quietly and leisurely dealing with the worm in accordance with its special meaning, as it does when there is no rival in the field, the chick darts at it, and bolts with it in accordance with the special meaning which its neighbour's presence, under such circumstances, has acquired. And this different behaviour carries with it a felt difference in the conscious situation—the interest of which is

centred in the companion. Or take the case of a herd of
cattle, which attacks a common enemy. The enemy is the
primary nucleus of the situation, but it is profoundly modified
by the presence of companions by which the behaviour of attack
is determined. The situation is social, and not merely individual, and a social interest suffuses it, and gives it a distinctive
character.

In this social interest probably arise the germs—but only the
germs—of the sense of personality. Some, indeed, go so far as
to urge that we learn to know ourselves only through knowing
others. The genetic order, so far as there is an order, is, they
say, not first the ego and then the alter, but first the mother
and companions and then through them the self. Or, to put
this point of view in a less questionable form, it is only
through the reaction of one on the other that the two are
differentiated. Be this as it may, it is only through the action
of environment on the organism, and the reaction of the organism on the environment in behaviour, that experience becomes
polarized into subject and object. Let it be clearly understood
that for the animal, in all probability, subject and object are
not clearly distinguished, and set over against each other in the
antithesis of thought. Only late in mental development are
the self and the world distinguished in subtle analysis as
different aspects of the common experience in which both
have their inseparable being. Animals, and perhaps the
majority of mankind, never trouble themselves about object
and subject as clearly realized products of conception and
reflective thought. For these concepts are exceedingly subtle.
And here, too, the external aspect of experience has the precedence, so far as there is precedence. A healthy lad from
the moment he gets up in the morning till the moment he
goes to bed, lives chiefly in the objective aspect of experience,
an aspect which is in us chiefly associated with the products in
consciousness of the leading senses of sight and hearing.
But the subjective aspect creeps in when he is hurt, when he
is hungry, when he is fatigued. He does not argue about the
matter, or formulate it in definite terms. He just dimly

feels that the interest has somehow shifted. Still more dimly does the animal feel that, apart from external interests which prompt nine-tenths of its behaviour, chiefly through the senses of sight, hearing, and smell, there are also matters in which the interest has somehow shifted to his own body. For the germ of self is essentially an embodied self. And perhaps the emotions, which ring through the system for some time after the external cause has been removed, serve in some degree to aid in this dimly felt shifting of interest.

Whatever may be the exact psychological nature of the emotions—and there has been much discussion of the question —it may be regarded as certain that they introduce into the conscious situation elements which contribute not a little to the energy of behaviour. They are important conditions to vigorous and sustained conation. And so closely interwoven are these elements with the whole situation in its impulsive aspect, that their disentanglement, in psychological analysis, is a matter of extreme difficulty. I have elsewhere * devoted some space to the consideration of the matter. I there follow Professor William James in regarding organic effects, other than motor sensations, as specially characteristic of the emotions in their primary genesis. The cold sweat, the dry mouth, the catch of the breath, the grip of the heart, the abdominal sinking, the blood-tingle or blood-stagnation—these and their like, in varied modes and degrees, characterize the emotions of fear, dread, anger, and so forth, when they rise to any pitch of intensity, and contribute largely to their sharpness and piquancy. These organic effects may be regarded as part of the private and individual business of the body ; but in ex- perience they closely coalesce with the motor effects through which the animal has to deal in practical behaviour with that which evokes the emotion.

On this view these organic states which contribute cha- racteristic elements in the emotional consciousness are due to afferent data from the vascular system and visceral organs, just as motor consciousness is due to afferent data from the

* " Habit and Instinct," ch. ix., p. 186.

parts concerned in overt behaviour. But, associated with emotional states, there are also certain motor reactions, which we speak of as their " expression "—so carefully discussed and elucidated by Darwin,—and these unquestionably contribute data to consciousness which coalesce with those afforded by the visceral and vascular elements. The whole is commonly suffused with feeling-tone, and the object which excites the emotion is a centre of pleasurable or painful interest. Representative elements, as experience develops, crowd into the conscious situation and render it more complex. And in addition to all this, there is, apart from the motor expression, the strenuous behaviour of flight or attack, or other mode of vigorous procedure which we commonly speak of as the outcome of the emotional state. The conscious situation, in the case of an enraged or scared animal actually behaving as such, is thus exceedingly complex. And it should be understood that in urging the importance of vascular and visceral elements, this complexity is nowise denied. What is suggested is that these elements are essential, and that they serve to characterize the distinctively emotional factor in the situation, that in any case they heighten the conative tendency.

Sufficient has now been said to indicate—but scarcely more than indicate—the importance of feeling-tone, interest, and emotion in determining the nature, character, and effective energy of the conscious situations which arise in the course of animal behaviour. They largely influence, and in part direct, the course of the conative tendency. But they also occur as its sequel. In animal, as in human life, the successful attainment of the end towards which conation sets is highly pleasurable. The equilibrium that is reached after instability, though it marks the close of present endeavour, leaves after-effects in consciousness in a sense of satisfaction which enters re-presentatively into later situations and helps to further more strenuous endeavour.

II.—PLAY

" There are two quite different popular ideas of play," says Professor Groos, in his admirable work on " The Play of Animals." " The first is that the animal (or man) begins to play when he feels particularly cheerful, healthy, and strong ; the second that the play of young animals serves to fit them for the tasks of later life." The former view, in which the latter may be included incidentally as a result, is closely associated with the names of Schiller,* who suggested it, and of Mr. Herbert Spencer,† who developed it. Mr. Wallaschek ‡ expresses the conception briefly and clearly when he says, " It is the surplus vigour in more highly developed organisms, exceeding what is required for immediate needs, in which play of all kinds takes its rise, manifesting itself by way of imitation or repetition of all those efforts and exertions which are essential to the maintenance of life." That surplus vigour is often a condition favouring the manifestation of play is probable enough, and seems to be supported by observation and experience ; but that it is likewise a condition favouring the chase, combat, mating, and much of the serious business of animal life seems equally unquestionable. Success in all these matters is largely determined by overflowing energy. In play, however, this surplus vigour finds vent when there is no serious occasion for its exercise. But, as Professor Groos says,§ " while simple overflow of energy explains quite well that the individual who finds himself in a condition of overflowing energy is ready to do something, it does not explain how it happens that all the individuals of a species manifest exactly the specific kind of play expression which prevails with their own species, but differs from every other." And if to this it be replied, that the specific kind is determined by repetition or

---

* " Letters on the Æsthetic Education of Mankind," xxvii.
† " Principles of Psychology," § 533.
‡ " On the Origin of Music," pp. 231, 232.
§ " The Play of Animals," Eng. trans., p. 12.

imitation of what we have called the serious business of animal life, Professor Groos's rejoinder is,* that "the conception of imitation here set forth—namely, as the repetition of serious activities to which the individual has himself become accustomed—cannot be applied directly to the primary phenomena of play—that is, to its first elementary manifestations" prior to any experience of these serious activities. The repetition (with a difference !) is in such cases not the re-enactment of what has been previously performed in full earnest by the individual, but rather the reappearance in the young of ancestral modes of procedure—in other words, its specific character is such because it is a piece of instinctive behaviour or arises from instinctive proclivities. And this is the central point of the interpretation elaborated with great skill and candour by Professor Groos. Play is instinctive ; and its biological value lies in the training it affords for the subsequent earnest of life.

Before leaving the surplus energy theory of play one more point made by Professor Groos may be mentioned. He contends that, though superabundant energy is a favouring condition of animal play (as it is, indeed, of all animal behaviour), still it is not a necessary condition. Animals often play when they are tired out. "Notice a kitten when a piece of paper blows past. Will not any observer confirm the statement that, just as an old cat must be tired to death or else already filled to satiety if it does not try to seize a mouse running near it, so will the kitten, too, spring after the moving object, even if it has been exercising for hours and its superfluous energies are entirely disposed of ? Or observe the play of young dogs when two of them have raced about the garden until they are forced to stop from sheer fatigue, and they lie on the ground panting, with tongues hanging out. Now one of them gets up, glances at his companion, and the irresistible power of his innate longing for the fray seizes him again. He approaches the other, sniffs lazily about him, and, though he is evidently only half inclined to obey the powerful impulse, attempts to seize

* *Op. cit.*, p. 7.

his leg. The one provoked yawns, and in a slow, tired kind of way puts himself on the defensive ; but gradually instinct conquers fatigue on him too, and in a few minutes both are tearing madly about in furious rivalry until want of breath puts an end to the game. And so it goes on with endless repetition, until we get the impression that the dog waits only long enough to collect the needed strength, not till superfluous energy urges him to activity." *

Coming now to Professor Groos's interpretation of play, we find in it, perhaps for the first time in the literature of the subject, adequate stress laid on its biological value. " The play of young animals," he says,† " has its origin in the fact that certain very important instincts appear at a time when the animal does not seriously need them. . . . Its utility consists in the practice and exercise it affords for some of the more important duties of life, inasmuch as selection [in the higher animals] tends to weaken the blind force of instinct, and aids more and more the development of independent intelligence as a substitute for it. At the moment when intelligence is sufficiently evolved to be more useful in the struggle for existence than the most perfect instinct, then will selection favour those individuals in whom the instincts in question appear earlier and in less elaborated forms—in forms that are merely for practice and exercise,—that is to say, it will favour those animals which play. . . . Animals cannot be said to play because they are young and frolicsome, but rather they have a period of youth in order to play ; for only by so doing can they supplement the insufficient hereditary endowment with individual experience, in view of the coming tasks of life."

Some stress is here laid on the fact that important instincts appear at a time when the animal does not seriously need them. It seems to imply the doctrine of what biologists term " acceleration "—which means the development of an organ or mode of behaviour at an earlier period in the descendants than that at which it appeared in the ancestors. Thus the adult fighting or hunting instinct of past generations appears in the young

* *Op. cit.*, p. 19.

† *Op. cit.*, pp. 75, 76.

to-day as a fighting-play or a hunting-play. It is open to question, however, whether either the instinctive behaviour or the conscious situation in the one case and the other is so nearly identical that the playful fight or hunt can fairly be called the same instinctive procedure as the serious combat or chase. We may hold, with Professor Groos, that the one is an invaluable preparation for the other without identifying them as the same behaviour under different conditions. Indeed the conditions are so different that the identification seems strained. The question may be left open, however, without impairing the value of Professor Groos's suggestion. And we may divide the preparatory behaviour in what is commonly called play under two heads : first, general preparation for varied modes of serious effort in after-life ; and secondly, special preparation for particular forms of this after-effort. Under the first will fall what Professor Groos terms experimentation and movement play, including what Dr. Stout, who fully realizes its importance, calls "manipulation " ; * under the second, such forms as hunting-play and fighting-play.

Nothing is more characteristic of the young of intelligent animals than the variety and persistency of their behaviour, their sensitiveness to stimuli of many different kinds, their restlessness of swiftly changing attention and response, with occasional pauses of continued effort in some special direction. Constantly on the alert, they exhibit in all its shifting phases behaviour which we interpret as indicating curiosity, inquisitiveness, love of mischief, destructiveness, and so forth. The facts are so familiar to every observer of young animals that it is unnecessary to give any detailed illustration. Watch a kitten in this stage of its development and carefully note its behaviour during half an hour ; the variety of effort, the *rôles* played by trial, failure, and success, the gain of skill and control over behaviour, will at once be evident. Or devote an equal space of time to observing young jays, magpies, or jackdaws. Every projecting piece of wire or bit of wood in their cage is pulled

* " Manual of Psychology," p. 327.

at this way and that way, from above, from below, from the side. Now one, then another, loose object is picked up and dropped, turned over, carried about, pulled at, hammered at, stuffed into this corner and into that, and experimented with in all possible ways. Then the wise bird goes to sleep, and wakes up again only to resume with new zest its persistent and varied efforts, by which it becomes acquainted with all the details of its environment. Watch young birds on the wing gaining their mastery of the air in flight, young seals tumbling in the water, young foals scampering and kicking up their heels in the meadows. A little observation, as occasion serves, a little attention to the progress towards an adequate experience of the meaning of things, towards more complete control, and increased nicety of behaviour, whether in reference to their surroundings, or in powers of finished locomotion, will serve to bring home what Professor Groos includes under experimentation and movement plays. He regards it all as play, since it seems to have no serious end, and is just a preparation for the sterner realities of adult life. And for human beings, whose work is so largely enforced, the freedom and evident joy of it all suggests the play which has acquired for us the meaning of relaxation from irksome effort, and glad abandonment to less constrained modes of behaviour. But in young animals such play is, after all, the serious business of their time of life. Its import for their future welfare can scarcely be overestimated.

And its import is in large degree psychological. If we watch a young puppy or kitten learning gradually to deal effectively with some difficulty in its extending environment, we see that it puts forth its efforts at first in a somewhat random and indefinite fashion. It is one of those animals in which intelligence has been evolved to supersede and become the more plastic substitute for instinct. The random and indefinite movements, are in detail reflex responses to stimuli. But whereas, in a piece of highly elaborated instinctive behaviour, such reflexes are grouped into a whole which is co-ordinated through inherited nervous mechanism ; in the

case of the acts of the puppy or kitten they have to be further
co-ordinated, or more elaborately grouped, through experience.
To act in one way some of the reflexes have to be checked as
redundant and not to the point : to act in another way other
reflexes have to be similarly checked ; and in a third way,
yet others.   But in all three some of the reflexes are utilized
to different ends.   Many conscious situations contain common
elements ; and this tends to give unity to the developing
experience.   But they contain also elements and groupings
which afford that diversity without which conscious behaviour
could not be accommodated to them.   So that we have here
the conditions under which what is technically termed " the
concomitant differentiation and integration of experience "
can proceed.

And if we speak of the instinct of experimentation we
must remember that what we are dealing with is rather an
innate tendency or instinctive propensity than a definite and
relatively clean-cut piece of instinctive behaviour.   It comprises a great number of inherited reflex acts, and may perhaps
be fairly called instinctive in detail.   But experimentation
must be regarded rather as the proximate end of a conative
tendency, or group of conative tendencies, whose ultimate
biological end is success in dealing with the environment in
the sterner struggle for existence during adult life.   The
tendency is inherited, and therefore falls under the head of
instinctive propensity.   But " experimentation " is a group-
term under which we comprise the general drift of varied
modes of behaviour, founded indeed on a congenital basis,
but receiving its stamp and character from what is acquired
in the course of the experience it provides.   It is essentially
a process whereby the conscious situations acquire what
Dr. Stout terms meaning; and is specially interesting as
affording an example of the way in which intelligence moulds
and refashions a number of disconnected reflex responses.   And
if, following Professor Groos, we call it play, it is a little
difficult to see how it can be brought in line with his statement *

* *Op. cit.*, pref., p. xx.

that "the play of youth depends on the fact that certain instincts, especially useful in preserving the species, appear before the animal seriously needs them." Does experimentation occur before it is needed in the economy of animal behaviour ? And might we not with equal truth say that the play of youth depends on the fact that certain acquired habits, especially useful in preserving the species, are gained before the animal seriously needs them ?

Passing now to those forms of play which afford more special preparation for particular forms of after-effort, under which fall such types as hunting-play and fighting-play, we may refer the reader to the copious examples so carefully collected by Professor Groos. The way in which a kitten pats a cork or a ball, making it roll and then pouncing upon it, is a characteristic example of animal play. Valuable as a preparation for dealing successfully with a mouse when occasion shall arise, this is a specialized form of experimentation ; and it is more obviously in line with the hunting-behaviour of later life than is general experimentation with any particular modes of future behaviour. Still it is essentially experimentation, with the instinctive propensity setting in more definite channels. Its value lies in the acquisition of skill under circumstances easier than those presented in the serious chase. So, too, in the case of the playful tussles of puppies or in that of the kitten, which not only shows playful fight to its brothers and sisters, but also to its mother, who responds by holding down the struggling and scratching little creature. Unquestionably, there is an instinctive propensity ; much of the detail, and some of the grouping, exhibit inherited reflexes due to special modes of stimulation. No doubt many of these responses occur in a similar but more emphatic way in a serious fight, and yet we may hesitate before committing ourselves to the theory of acceleration. It is at least equally probable that play as preparatory behaviour differs in biological detail (as it almost certainly does in emotional attributes) from the earnest of after-life, and that it was evolved directly as a preparation, as a means of experimentation through which certain essential modes of skill were acquired,—those animals in which the preparatory play propensity was not inherited in due force and requisite amount being worsted in the combats of later life, and eliminated in the struggle for existence. For, in the preparatory tussles and squabbles and playful fights of young animals, experience is gained without serious risk to life and limb.

The modifications of Professor Groos's biological interpretation of play which we would suggest are so slight that we may be said to accept it almost unreservedly. The play of youth, we may urge, depends on instinctive propensities to experimentation in varied ways, some of more general and others of more special import ; and the value of such experimentation lies in the fact that it is a means of acquiring, under circumstances more easy and less dangerous than those of sterner life, experience and skill for future use. In a word, play depends on instinctive propensities of value in education.

Passing now to a brief consideration of the feelings and emotions which we may suppose to accompany play, we may place first those which characterize, from this point of view, general experimentation. We have here rapidly varying situations charged with conative impulse, the satisfaction of which must bring pleasure—the occasional thwarting of which is probably toned with the opposite—the latter serving, through contrast, to enhance the satisfaction of ultimate success. Both pleasure and its antithetical state of feeling are primarily matters of the conscious situation as a whole, and even in ourselves are difficult to distribute in analysis. But assuredly no small share of the total product must be assigned to the successful behaviour which consummates the conative tendency. Indeed, it is the thwarting of free action which is the source of much of the discomfort of the young. Unimpeded and vigorous behaviour also brings with it secondary effects in organic processes—fuller heart-beat, freer circulation, deeper respiration, better digestion, firmer muscular tone, and so forth— which have a marked effect on the conscious situation, and aid in producing that emotional tone which cannot, perhaps, be named in better terms than "good spirits" and the joy of existence, so forcibly suggested during the free play of youth. On the other hand, there is no more piteous sight than that afforded by the young animal, "cabined, cribbed, confined," suffering from *ennui* and depression—all its organic processes sluggish and craving to be quickened into the natural vigour of life, not creeping slowly through the veins, but coursing at full flood.

In the psychological aspect of play Dr. Groos assigns perhaps the first place to pleasure in the possession of power, or, as Preyer phrases it, pleasure in being a cause. We must be careful, however, lest in using such expressions we seem to imply that animals—even quite young animals—are capable of entertaining ideas which belong to a much later stage of mental development. Speaking of "joy in ability or power," Professor Groos says,* "This feeling is first a conscious presentation to ourselves of our personality as it is emphasized by play. . . . But it is more than this; it is also delight in the control we have over our bodies and over external objects. Experimentation in its simple as well as its more complicated forms is, apart from its effect on physical development, educative in that it helps in the formation of causal associations. . . . The young bear that plays in the water, the dog that tears a paper into scraps, the ape that delights in producing new and uncouth sounds, the sparrow that exercises its voice, the parrot that smashes his feeding trough—all experience the pleasure in energetic activity, which is, at the same time, joy in being able to accomplish something." But those who agree with Dr. Stout, as I do without hesitation, in denying personality (save in a very embryonic condition) and the conception of causation to animals in the perceptual stage of mental evolution, though they may find in Dr. Groos's contention a central core of truth, will be unable fully to accept his manner of presenting it. "Any single train of perceptual activity," † says Dr. Stout, "has internal unity and

* *Op. cit.*, p. 290.
† "Manual of Psychology," p. 266.

continuity. But where conscious life is mainly perceptual, the
several trains of activity are relatively isolated and disconnected with each other. They do not unite to form a
continuous system, such as is implied in the conception of a
person. We must deny personality to animals." To this I
would merely add that, even where perceptual continuity in
animals reaches its maximum, it is not reflectively grasped as
a whole, and the ideal construction of the personal ego is not
conceived as antithetical to the impersonal world of objects.
With what Dr. Stout says about causality I am in complete
agreement. "We must notice," * he urges, "the essential
difference which separates the merely perceptual category from
that of ideational and conceptual thought. The perceptual
category is always purely and immediately practical in its
operation. It is a constitutive form of thought only because
it is a constitutive form of action. The question 'Why?' has
no existence for the merely perceptual consciousness. It does
not and cannot inquire how it is that a certain cause produces
a certain effect. It does not and cannot endeavour to *explain*,
to analyze conditions so as to present a cause as a *reason*. It
does not compare different modes of procedure or different
groups of circumstances, so as to contradistinguish the precise
points in which they agree from those in which they disagree,
and in this way to explain why a certain result should follow
in one case and a different result in another case. Causality
in this sense can only exist for the ideational consciousness,
and the development of the ideational consciousness in this
direction is a development of conceptual thinking—of generalization."

Wherein, then, lies the central core of truth in Professor
Groos's contention? In the satisfaction that arises from the
success of any conative activity. We see that the animal
striving and doing falls within our conception of a cause, in
the scientific sense of the word,—a relatively constant and
continuous antecedent of diverse sequent effects. We infer
that pleasure accompanies the satisfaction of the multifarious

* *Op. cit.*, p. 314.

conative impulses. The pleasure is the animal's ; the conception
of causality and of self as a continuous person, still the same
amid diversity of conscious situations, is ours. If we bear this
in mind there can be no objection to our attributing to
animals joy in ability or power. It is the pleasure derived
from that successful conation whereby animals fall into the
category of causes within the scheme of our rational thought.

In fighting-play and hunting-play, too, there arise in
more specific forms the pleasures of successful conation with
the antithetical feelings accompanying thwarted conation.
And these are distinguished from earnest, partly because the
companion or the inanimate substitute for prey is the centre
of a different situation from that afforded by an enemy or the
natural object of the chase ; partly by the absence of certain
insistent emotional states which characterize earnest and
the serious business of life. In fighting, this is anger. And
we often see the tendency of this to arise in the midst of
fighting-plays, and at once say that it becomes serious and
passes into fighting in earnest. Indeed, some tinge of earnest,
with its fuller emotional tone, forms part of the preparation
for future life, and so far falls within the definition Professor
Groos gives of play. From which we may see that play is not
easily marked off from other forms of conation.

Brief reference to the element of "make-believe," which
Dr. Groos assigns to the higher forms of play, may be reserved
for our fourth section ; and some further discussion of its
psychological aspect to the concluding chapter.

III.—COURTSHIP

We have seen that Professor Groos regards play as the
practice and preparation for the serious business of animal
life. Founded on instinctive tendencies, it has its biological
value in the acquisition of practical acquaintance with the
environment, and of skill in dealing with it effectually. It is
an education in behaviour of the utmost service in view of the
struggle for existence. It is full of the pleasure derived from

the satisfaction of innate impulse, the success of conative effort, and the diffused sense of well-being which accompanies a life of action, free and unrestrained. This freedom and gladness lead us to call it play ; but we must not draw the inference that the playful animal knows that it is playing, or forms any conception of the antithesis between work and play, which is a product of late development.

In laying stress on the biological value of certain modes of behaviour which we thus call play, a value which lies in the practice and preparation they afford for life's more earnest work, Professor Groos deserves our hearty thanks. Nor need our thanks be less hearty if we find that he has in some degree been anticipated by Darwin ; for he has elaborated with systematic care what Darwin suggested incidentally. " Nothing is more common," said Darwin,* " than for animals to take pleasure in practising whatever instinct they follow at other times for some real good. How often do we see birds which fly easily, gliding and sailing through the air, obviously for pleasure ! The cat plays with the captured mouse, and the cormorant with the captured fish. The weaver bird, when confined in a cage, amuses itself by neatly weaving blades of grass between the wires of its cage. Birds which habitually fight during the breeding season are generally ready to fight at all times ; and the males of the capercailzie sometimes hold their *Balzen*, or *leks*, at the usual place of assemblage during the autumn. Hence it is not at all surprising that male birds should continue singing for their own amusement after the season of courtship is over."

In the behaviour of courtship we have what is essentially part of the serious business of animal life. And in including it under the heading of " Love Plays," Professor Groos may seem to be forgetful of his own definition of play. He is, however, too clear a thinker not to see, and too honest an exponent not to say, that much of the emotional behaviour commonly regarded as courtship falls outside his main thesis " in being, strictly speaking, not mere practice preparatory to

* " Descent of Man," vol. ii., p. 60, 2nd edit. 1885.

the exercise of an instinct, but rather its actual working." *
But behaviour of a somewhat similar kind is seen in young
animals before the time of mating has arrived, and is exemplified both in young and adults under circumstances different
from those which distinguish what we may term the pairing
situation. This, at any rate, may be regarded as a form of
experimentation and practice in the arts of courtship. On
different grounds does Professor Groos attempt to justify the
inclusion of actual courtship under the head of play. For it
may also, he thinks, even at the time of its serious exercise, be
to some extent artful, involving " make believe," and therefore playful in a somewhat different and more subtle sense ;
but a brief reference to " make believe " we may reserve for
our next section.

There can be no question that special modes of behaviour
often characterize the pairing situation, and that these not
only exemplify an instinctive tendency, but from their constancy and relative definiteness constitute types of instinctive
behaviour. They would form parts of any definition of a
species founded not on structure but on behaviour. And if
animals have feelings and emotions at all—if they are not
Cartesian automata, which merely *seem* to be guided in their
actions by consciousness—there can be but little question that
the behaviour which characterizes the sexual situation is unusually charged with feeling-tone, and accompanied by all
those adjuncts which distinguish an emotional state, broadly
considered. This matter is of no little importance in our
interpretation of the phenomena described as courtship. Do
the accompanying feeling-tone and the state of emotional
exaltation influence the behaviour, or would it run a similar
course in the absence of any such accompaniments ? If, as we
can scarcely doubt, the consciousness attending the situation
does profoundly influence the behaviour, the further question
arises—Is this influence mainly the result of the presence and
behaviour of an individual of the opposite sex ? To this, again,
we must answer that, so far as we can learn by observation,

* "The Play of Animals," Eng. trans., p. 229.

behaviour unquestionably is determined by such influence in the serious business of courtship. And then the further question arises—Is it a matter of indifference what the appearance and behaviour of the individual of the opposite sex may be? Are A., B., C., and the rest of the male alphabet, precisely alike in stimulating in a similar manner, and to a similar degree, the sexual impulse of the female? If we admit any differential influence, and if this influence takes effect in the sexual union to which courtship is preparatory, we so far admit the efficacy of that which Darwin termed sexual selection.

Let us, however, before proceeding with general considerations, present one or two examples of the facts which observation has furnished with regard to specialized modes of behaviour at the time of pairing. Speaking of American night-hawks, Audubon says, "Their manner of flying is a good deal modified at the love season. The male employs the most wonderful evolutions to give expression to his feelings, conducting them with the greatest rapidity and agility in sight of his chosen mate, or to put to rout a rival. He often rises to a height of a hundred metres and more, and his cries become louder and louder as he mounts; then he plunges downward with a slanting direction, with wings half open, and so rapidly· that it seems inevitable that he should be dashed to pieces on the ground. But at the right moment, sometimes when only a few inches from it, he spreads his wings and tail, and, turning, soars upward once more." * Mr. Strange, quoted by Darwin,† says of the satin bower-bird, "at times the male will chase the female all over the aviary, then go to the bower, pick up a gay feather or a large leaf, utter a curious kind of note, set all his feathers erect, run round the bower and become so excited that his eyes appear ready to start from his head; he continues opening first one wing, and then the other, uttering a low, whistling note, and, like the domestic cock, seems to be picking up something from the ground, until at last the female goes quietly towards him."

* Quoted by Groos, *op. cit.*, p. 259.
† "Descent of Man," vol. ii., p. 77.

Darwin describes how, in the Argus pheasant,[*] "the immensely developed secondary wing-feathers are confined to the male, and each is ornamented with a row of from twenty to twenty-three ocelli, each above an inch in diameter. These beautiful ornaments are hidden until the male shows himself before the female. He then erects his tail, and expands his wing-feathers into a great, almost upright, circular fan or shield, which is carried in front of the body. The ocelli are so shaded that, as the Duke of Argyll remarks, they stand out like balls lying loosely within sockets. But when I looked at a specimen in the British Museum, which is mounted with the wings expanded and trailing downward, I was," adds Darwin, " greatly disappointed, for the ocelli appeared flat or even concave. Mr. Gould, however, soon made the case clear to me, for he held the feathers erect, in the position in which they would naturally be displayed, and now, from the light shining on them from above, each ocellus at once resembled the ornament called a ball and socket." The primary wing-feathers are scarcely, if at all, inferior in beauty to the secondaries, though the markings are quite different, the chief ornament being a space parallel to the dark-blue shaft, which in outline forms a perfect second feather lying within the true feather. "Now the secondary and primary wing-feathers are not at all displayed, and the ball and socket ornaments are not exhibited in full perfection, until the male assumes the attitude of courtship."

It is unnecessary to describe the song of birds which is generally, but not always, at its best during the period of pairing. Bechstein, who kept birds during his whole life, and studied them with care, asserts that " the female canary always chooses the best singer, and that in a state of nature the female finch selects that male out of a hundred whose notes please her most." [†]

Thus we are led back to sexual selection. If we are satisfied that the males of certain species do as a matter of fact behave

[*] Darwin, *op. cit.*, p. 99.

[†] Quoted by Darwin, " Descent of Man," vol. ii., p. 58.

in specific and distinguishable ways at the breeding season, and in presence of their would-be mates, the question is, What, if any, is the biological value of such behaviour? What has fostered and guided it in the course of its evolution? From the case of the Argus pheasant, which is only a sample of the large class of cases in which the male has special adornments, we see that the behaviour has often direct relation to the display of such plumage, or, in some apes, of coloured surfaces, so that behaviour and ornamentation must be taken together. The essence of Darwin's contention is, that the adornments and behaviour give rise to a situation through which the female is stimulated or excited to accept the male; that the male in which they are best developed gives rise to the most effective situation, produces most excitement, and therefore has the best chance of acceptance, being " unconsciously preferred ; " * and that he thus begets offspring which inherit his adornments and modes of behaviour, such inheritance being, however, confined to the males. Thus sexual selection takes effect through preferential mating, whereby certain hereditary traits are transmitted and become racial characteristics. And this is brought about through an appeal to consciousness, and seems to involve choice—generally that of the female.

Now, as I have elsewhere urged,† the hypothesis of sexual selection has often been placed in a false light by the introduction of the unnecessary supposition that the hen bird, for example, must possess a standard or ideal of æsthetic value, and that she selects that singer which comes nearest to her conception of what a songster should be. Darwin occasionally expressed himself unguardedly in the matter ; he says, for example, that the female appreciates the display of the male, and places to her credit a taste for the beautiful.‡ But he also distinctly states that " it is not probable that she consciously deliberates ; she is most excited or attracted by the most beautiful, or melodious, or gallant males." § This is all that is really necessary for the theory of sexual selection. The hen

* " Descent of Man," vol. ii., p. 56.   † " Habit and Instinct," p. 217.
‡ " Descent of Man," vol. ii., p. 251.   § *Op. cit.*, p. 137.

accepts that mate which by his song or otherwise excites in
sufficient degree the pairing impulse ; if others fail to excite
this impulse, they are not accepted.   Even Mr. A. R. Wallace,
who rejects the theory save in a very subordinate form, says *
that " it may be admitted, as highly probable, that the female
is pleased or excited by the display," and speaks † of a possible
choice of " the most vigorous, defiant, and mettlesome male,"
giving, moreover,‡ several telling examples of preferential
mating.   Stripped of all its unnecessary æsthetic surplusage,
at any rate so far as this implies an æsthetic ideal, or æsthetic
motive, the hypothesis of sexual selection suggests that the
accepted mate is the one which adequately evokes the pairing
impulse.

Here Dr. Groos makes an interesting and important con-
tribution to the subject.   He lays stress on the coyness and
reluctance of the female, and illustrates it by examples derived
from observation.   "Thus the female cuckoo answers the call
of her mate with an alluring laugh that excites him to the
utmost, but it is long before she gives herself up to him.
A mad chase through tree-tops ensues, during which she constantly incites him with that mocking call, till the poor fellow
is fairly driven crazy.   The female kingfisher often torments
her devoted lover for half a day, coming and calling him, and
then taking to flight.   But she never lets him out of her sight
the while, looking back as she flies and measuring her speed,
and wheeling back when he suddenly gives up the pursuit.
The bower-bird leads her mate a chase up and down their
skilfully built pleasure-house, and many other birds behave in
a similar way.   The male must exercise all his arts before her
reluctance is overcome.   She leads him on from limb to limb,
from tree to tree, until it seems that the tantalizing change
from allurement to resistance must include an element of
mischievous playfulness." §

Professor Groos regards the instinctive coyness of the
female as the most efficient means of preventing the too early

* " Darwinism," p. 285.   † *Op. cit.*, p. 293.
‡ *Op. cit.*, p. 172.   § *Op. cit.*, pp. 285, 286.

and too frequent yielding to sexual impulse.* He thinks it probable that "in order to preserve the species the discharge of the sexual function must be rendered difficult, since the impulse to it is so powerful that, without some such arrest, it might easily become prejudicial to that end. This same strength of impulse is," he adds,† "itself necessary to the preservation of the species ; but, on the other hand, dams must be opposed to the impetuous stream, lest the impulse expend itself before it is made effectual, or the mothers of the race be robbed of their strength, to the detriment of their offspring." It has its origin in the general fact that, before any important motor discharge takes place, there is apt to be a preparatory and gradually increasing excitement. But this is specially emphasized in association with the sexual impulse. As Professor Zeigler wrote, in a private communication to Dr. Groos,‡ " Among all animals a highly excited condition of the nervous system is necessary for the act of pairing, and consequently we find an exciting playful prelude very generally indulged in."

Courtship may thus be regarded from the physiological point of view as a means of producing the requisite amount of pairing-hunger ; of stimulating the whole system and facilitating general and special vascular changes ; of creating that state of profound and explosive instability which has for its psychological concomitant or antecedent an imperious and irresistible craving. This not only overcomes the coyness of the female, but generates and strengthens the ardour of the male—a point on which, perhaps, Professor Groos does not lay sufficient stress. For the process is reciprocal ; and though the male leads in the ardour of courtship, yet this ardour constantly grows till at last it overcomes the barriers of reluctance. Courtship is thus the strong and steady bending of the bow that the arrow may find its mark in a biological end of the utmost importance in the survival of a healthy and vigorous race.

The coyness and reluctance of the female afford the

* *Op. cit.*, p. 283.   † *Op. cit.*, p. 243.
‡ *Op. cit.*, p. 242.

conditions under which the bow is bent to the full. But they also afford the conditions of the apparent act of choice. This takes the place, on the perceptual plane of mental development, of that deliberation which precedes the higher act of choice on the ideational plane. For psychology, as well as for biology, then, Dr. Groos's suggestion is a welcome and helpful one. Both upholders of sexual selection and critics of that hypothesis, have been apt to regard the choice of a mate in animals from too anthropomorphic a point of view—to look upon it as the outcome of rational deliberation, of weighing in the æsthetic balance the relative attractiveness of this suitor and of that, and of reaching a definite conclusion that the one is to be accepted because he behaves or is adorned in such and such a way, while the other is to be rejected because he falls below all reasonable standards of requirement. The choice exercised by the female, if so we term it, is far simpler and more naïve. Indeed, Professor Groos goes so far as to say that on his hypothesis the element of choice is altogether abolished. " It is the instinctive coyness of the female," he says,* "that necessitates all the arts of courtship, and the probability is that seldom or never does the female exert any choice. She is not awarder of the prize, but rather a hunted creature. So, just as the beast of prey has special instincts for finding his prey, the ardent male must have special instincts for subduing female reluctance. According to this theory, there is choice only in the sense that the hare finally succumbs to the best hound, which is as much as to say that the phenomena of courtship are referred at once to natural selection." He reaches this conclusion, however, by gradual stages. He first urges † that " it would be absurd to affirm that all bird songs originate in a conscious æsthetic and critical act of judgment on the part of the female. A conscious choice either of the most beautiful or the loudest singer is certainly not the rule, and probably never occurs at all. But," he adds, "is it not still a choice, though unconscious, when the female turns to the singer whose voice, whether from strength or modulation,

* *Op. cit.*, pref., p. xxii.   † *Op. cit.*, p. 240.

proves most attractive ? Even if the song is primarily a means of recognition or an invitation from the male, still the psychological effect must be that the female follows the songster that excites her most, and so exerts a kind of unconscious selection."

The phrase " unconscious choice " is, however, somewhat unsatisfactory, especially when we remember that it is used to indicate the result of a direct appeal to the conscious situation. If, however, we say that it is perceptual choice arising from impulse as distinguished from ideational choice due to motive and volition, we see that the distinction is in line with that which we have drawn again and again, and which Dr. Stout so well emphasizes in his " Manual of Psychology." But we have also drawn the distinction between instinctive behaviour prior to individual experience, and intelligent behaviour the result of such experience. Under which of these classes does the behaviour of the female during courtship fall ? Professor Groos, in the further development of his hypothesis, seems to place it in the instinctive category. " Instead of a conscious or unconscious choice, of which we know nothing certain," he says,[*] " we have the need of overcoming instinctive coyness in the female, a fact familiar enough, but hitherto not sufficiently accounted for. Then the question no longer is, which among many males will be chosen by the female, but which one has the qualities that can overcome the reluctance of the female whom he woes. Sexual selection would then become a special case of natural selection."

I am unable to follow Professor Groos in this view, which I find it rather difficult to reconcile with such statements as that already quoted, in which he says that the female's " tantalizing change from allurement to resistance seems to include an element of a mischievous playfulness." It is more probable that instinctive coyness and reluctance afford the conditions under which experience of the pleasures of courtship may be gained. It is said that a flirt, when taken to task for her conduct at ball and picnic, justified it by asking demurely

* *Op. cit.*, p. 244.

how else she was to gain that wide experience of men which was absolutely necessary to guide her in exercising a wise and becoming choice. Let us hope that when the fateful time arrived she acted with due deliberation. Now the coquetry of birds affords the opportunity of gaining just such experience in the light of which a perceptual choice may be made.

Let us remember that courtship is, as Darwin said, "a prolonged affair," and that coyness is a means to its prolongation. And let us remember that in simple cases, as also in more complex matters, the intelligent exercise of choice depends upon what Dr. Stout terms the acquisition of meaning. "The chicken does not, at first," he says,* "distinguish between what is edible and what is not. This it has to learn by experience. It will at the outset peck at and seize all worms and caterpillars indiscriminately. There is a particular caterpillar called the cinnabar caterpillar. When this is first presented to the chicken it is pecked at and seized, like other similar objects. But as soon as it is fairly seized it is dropped in disgust. When next the chicken sees the caterpillar, it looks at it suspiciously, and refrains from pecking. Now, what has happened in this case? The sight of the cinnabar caterpillar re-excites the total disposition left behind by the previous experience of pecking at it, seizing it, and ejecting it in disgust. Thus the effect of these experiences [what I have termed the conscious situation] is revived. The sight of the cinnabar caterpillar has acquired a *meaning*." Take now the case of a coy hen bird, to whom several males pay court. The sight of this one, behaving after his kind, excites in small degree the sexual impulse and emotions. Her heart beats but little the faster for all his antics, her respiratory rhythm is scarcely affected, her feathers, like her feelings, remain comparatively unruffled. He has acquired meaning from the reaction to his presence ; it is not, however, a very attractive meaning. But that other, perhaps from mere persistency, perhaps because he is more " vigorous, defiant, and mettlesome " (she, at any rate, certainly knows not why), deeply

* "Manual of Psychology," p. 85.

stirs her organic being, sets her all aglow, and breaks down the barriers of her coyness. And this he does because he is the centre of a conscious situation which has acquired, through her experience of his presence, a meaning and an interest that are at last irresistibly attractive. It is a choice from impulse, not the result of deliberation ; but it is a choice which is determined by the emotional meaning of the conscious situation. And it is the reiterated revival of the associated emotional elements which generates an impulse sufficiently strong to overcome her instinctive coyness and reluctance.

And this coyness is the natural correlative of the ardour of the male, an ardour increased by his courtship antics. If the female yielded readily and at once, the behaviour of courtship would never have been evolved. Superabundant vigour in the male is, no doubt, a favourable condition of courtship, as it is of play ; but neither is it a *sine quâ non*, nor in any case does it, or can it, afford any guidance of behaviour into just those specific channels in which we find it setting during the breeding season. If sexual selection be not a *vera causa* of the specific direction, we have at present no other hypothesis which in any degree fits the facts. And to the criticism of those who, like Mr. W. H. Hudson, urge that dance and song, and aërial evolutions in birds, are seen at times when the immediate business of courtship forms no part of the situation, Professor Groos's theory of play affords a sufficient answer. If courtship, whose biological end is of such supreme importance, forms a central feature in the serious business of animal life ; and if play is the preparation and practice for behaviour of biological importance ; we should expect to find manifestations (with an emotional difference, and no doubt many differences in detail) of all those actions, the due performance of which, in the supreme hour of courtship, will alone enable the adequately prepared and wellpractised male to overcome the reluctance of the female, and beget offspring to transmit his instinctive and emotional tendencies.

IV.—Animal " Æsthetics " and " Ethics "

In this section we shall consider some types of behaviour which suggest situations that contain the germs of æsthetics and ethics, with a view to determining, so far as possible, the principles on which they should be interpreted. This is a peculiarly difficult subject; for we are endeavouring to get behind the behaviour, and to infer the mental conditions which accompany it, and through which it assumes its distinctive character. The difficulty is twofold : first, because, as Dr. Stout puts it,* " human language is especially constructed to describe the mental states of human beings, and this means that it is especially constructed so as to mislead us when we attempt to describe the workings of minds that differ in any great degree from the human ; " and secondly, because, to quote the same careful thinker,† " the besetting snare of the psychologist is the tendency to assume that an act or attitude which in himself would be the natural manifestation of a certain mental process must, therefore, have the same meaning in the case of another. The fallacy lies in taking this or that isolated action apart from the totality of conditions under which it appears. It is particularly seductive when the animal mind is the subject of inquiry."

We must, therefore, base our method of procedure on some definite principle. The canon of interpretation which I have elsewhere suggested ‡ is, that we should not interpret animal behaviour as the outcome of higher mental processes, if it can be fairly explained as due to the operation of those which stand lower in the psychological scale of development. To this it may be added—lest the range of the principle be misunderstood —that the canon by no means excludes the interpretation of a particular act as the outcome of the higher mental processes, if we already have independent evidence of their occurrence in

* " Manual of Psychology," p. 23.
† *Op. cit.*, p. 22.
‡ " Introduction to Comparative Psychology," p. 53.

the agent.   Now, the conclusion to which we are led by direct experiment and a critical study of the actions of animals whose life-history is known to us is, that most of their behaviour —perhaps all—is due to what Dr. Stout terms the perceptual, as opposed to the ideational, exercise of cognition.   Their behaviour can be explained without having recourse to the hypothesis that they reflect, and attain to ideal schemes as the result of abstraction and generalization consciously directed to this end.   Rather than repeat what I have already said, I will quote Dr. Stout's summary of the position to which he, too, has been led.   "The vast interval," he says,* "which separates human achievements, so far as they depend on human intelligence, from animal achievements, so far as they depend on animal intelligence, is connected with the distinction between perceptual and ideational process.   Animal activities are either purely perceptual, or, in so far as they involve ideas, these ideas serve only to prompt and guide an action in its actual execution.   On the other hand, man constructs ' in his head,' by means of trains of ideas, schemes of action before he begins to carry them out.   He is thus capable of overcoming difficulties in advance.   He can cross a bridge before he comes to it."

It has already been stated that in the intelligent behaviour of animals under man's teaching he is the rational agent, they his willing slaves.   This may be here again illustrated to enforce the distinction drawn by Dr. Stout in the above passage.   Those who have seen a shepherd's dog working sheep on a moorland fell, and have taken the trouble to ascertain how the results he sees have been attained, will appreciate, on the one hand, how well the dog knows and responds to the signals of his master, and, on the other hand, how completely all initiation is in the master's mind, not that of the keenly intelligent dog.   Those who merely witness such a performance without inquiry or investigation will probably misunderstand the whole matter.   In the north of England competitions are not uncommon where, say, three sheep have to be driven over a definite course, between certain posts and

* "Manual of Psychology," p. 266.

round others, through narrow passages and into a fold—all within a certain time limit. At such a competition success depends on two things : first, the training of the dog to respond at once to some six or eight whistle-signals, often accompanied by gestures and movements of a stick ; and secondly, the judgment of the shepherd. The signals, given in different whistle-tones and inflections, have for the dog meaning, such as drive straight on, from this side, from that, stop, lie down, creep, and so forth. The dog's whole business is to obey these signals. And the instant response of a well-trained dog is admirable. But in the whole proceeding he is merely the executant of his master's orders. He originates no important step. And if you listen to the criticisms by other shepherds during a competition you will find that they are mainly passed on the judgment shown by the master, and only in palpable failures in obedience on the behaviour of the dog. The intelligent animal is what he is trained to be—one whose natural powers are under the complete control of his master with whom the whole plan of action lies.

Since, then, in the cognitional field we find no independent evidence of the higher processes, we are bound, in accordance with our canon, to interpret emotional situations on similar principles, unless we find in them outstanding facts which cannot be explained in this way.

In considering the pairing situation we urged that the framing of an ideal of beauty to which a given suitor approaches, or from which he falls short, is unnecessary for the interpretation of the facts. We should not in strictness, therefore, speak of " an appreciation of beauty " or " a taste for the beautiful " in birds, since such expressions almost inevitably imply that these creatures have reached some conception of beauty as distinguished from and contrasted with ugliness. At the same time the hen certainly appears to enjoy the situation of which the plumed cock, attitudinizing thus, forms the centre of interest—through which he acquires meaning. Although, therefore, there is probably no ideal or standard of beauty, there are afforded the data in experience from which, were the bird capable of reflection, such an ideal might, in ideational sublimation, be derived. Before comparison, abstraction, and generalization can be applied, in the reflective laboratory of thought, there must be suitable experiences to form the raw material on which these rational processes can be exercised. Long ere, in the course of mental evolution, the correlative conceptions implied in the phrase " beautiful or ugly " had taken definite form, perceptual situations must have arisen, where, by direct appeal to the senses, by the diffused effects of stimulation and their accompanying feeling-tone, and by the natural satisfaction of mere impulse, the foundations were laid of that appreciation of the beautiful which forms the reflective superstructure we build upon them. Indeed, the pleasure and satisfaction attending particular situations, as they severally arise, appear to contain the perceptual germs of what in later development becomes aesthetic appreciation.

The bird which, having completed its nest, eyes it with apparent satisfaction, may well have the germs of that which, when rendered schematic in our thought, we call taste. Dr. Gould, indeed, states that certain humming-birds decorate their nests " with the utmost taste," weaving into their structure beautiful pieces of lichen. And the gardener bower-bird collects in front of its bower flowers and fruits of bright and varied colours. What meaning these carry in the conscious situation we do not know ; we can only suppose that they incidentally contribute to the heightening of the sexual impulse, and have been evolved as a means of stimulation to the biological end towards which sexual selection is unconsciously directed. For it is probable that all the situations with which pleasure and satisfaction are in high degree associated are, in primary origin, closely connected with behaviour directed, through natural or sexual selection, to some definite biological end, or, in brief, with behaviour of biological value. And it is, perhaps, not improbable that the states of consciousness most highly toned with strong emotion have their origin in those situations which arise amid the pairing, parental, and companiable relations of animal life.

We have already said that the companion, as the nucleus of a situation, is a thing which reacts in altogether special ways, so that it becomes differentiated from other things as something the meaning of which, and the interest in which, are *sui generis* and unique in type. It becomes the centre of emotional situations, which we ascribe to rivalry, emulation, jealousy, and so forth. And we have also drawn attention to the view that the genetic order, so far as there is an order, is not first the ego and then the alter, but first the mother and companions and then through them the self. We learn to know ourselves only through knowing others. We must now ask the question—a question which must be answered before we can touch on the possible ethics of animals—how far, and in what sense, the social animal regards others as of like nature to itself, and capable also of like feelings and emotions. Stated in this form we must, I think, answer the question in the negative. The expression, "of like nature to itself," implies that the self has already taken more or less definite form, and that the animal infers that, since the alter behaves and reacts in like manner to the ego, it also is an ego. This is distinctly an act of reasoning. As Clifford phrased it, the companion becomes an *eject*. We can never by direct experience become acquainted with the feelings of others, but we can endow them ejectively with personality analogous to our own.

But, though it is exceedingly doubtful whether any animal can regard its companion as an "eject," may there not be a perceptual anticipation of the ideational process that comes with later-developed reflection? A decade ago I gave the following answer to this question: "For myself, I cannot doubt that animals project into each other the shadows of the feelings of which they are themselves conscious." * Professor Mark Baldwin speaks of the stage at which this takes place, as the "projective stage" of development. "Now, in the fact," he says,† "of herding, common life and arrangements

* "Animal Life and Intelligence," p. 340.
† "Mutual Development in the Child and the Race," p. 19.

for the protection of the herd, animal societies of various kinds, animal division of labour, etc.,—whatever be the origin of it,— we have what seems to be such an epoch in animal life. These creatures show a real recognition of one individual by another, and a real community of life and reaction, which is quite different from the individualism of purely sensational and unsocial consciousness. And yet it is just as different from the reflective organization of human society, in which the self-consciousness and personal volition of the individual play the most important *rôle*. I see no way of accounting for the gregarious instinct anywhere, except on the assumption of such a projective epoch of animal consciousness."

Now, in endeavouring to realize how the situation feels to an animal in this projective stage, the first difficulty we encounter is that of divesting ourselves of those products of reflection which characterize our own mental situation ; and to avoid what Dr. Stout, in the passage above quoted,* terms the psychologist's besetting snare. The second difficulty is to grasp that, in experience, subject and object are inseparable, however clearly we may learn to perceive that they are dis- tinguishable aspects of that experience. If the subject is eventually regarded as that which experiences, and the object as that which is experienced, it is surely obvious that each is necessary to the other. But, before these different aspects are clearly distinguished, there is, in the perceptual stage of mental development, what we may term a distribution of the items of experience among the centres of interest.

In illustration of the kind of distribution which we may suppose to come naturally to an animal, in what Professor Baldwin terms the projective epoch, let us take three animal situations : first, a chick pecks at a soldier-beetle, and finds it nauseous ; secondly, a hen-bird hears the joyous song of her mate ; thirdly, a puppy in play bites its companion, and receives a painful nip in return. Each of these constitutes an experience-situation ; assuming that the results of the expe- rience are distributed, how may we suppose them to be allocated ?

* *Supra*, p. 270.

In the first case, the soldier-beetle is the centre of interest in the situation. As the situation develops, the element of nauseousness is introduced. As Dr. Stout puts it, this is what gives the soldier-beetle meaning. Can it be doubted that, if there be any distribution, the nauseousness, though it is altogether what we have learnt to call a subjective affection, attaches itself to the soldier-beetle? The plain man, unsophisticated by Berkeleyan discussion, says simply, in such cases, " The thing is nauseous." And this probably indicates the naïve and primitive distribution. Turning now to our second example, when the hen hears the courtship song the mate is the centre of a situation suffused with pleasurable feeling. How is the joyousness, again essentially subjective for our later thought, distributed? Surely, if at all, on the mate who forms the centre of interest. This it is which gives him meaning. The joy of the hearer is projected on to the singer. Not entirely, perhaps ; the hen literally, on Professor James's theory of the emotions, feels her heart-beat quickened by his presence, and the delightful ruffling of her feathers. But our aim is not to deny that the germs of the subjective arise in the midst of such situations, but to contend that some at least of the joyous character of the situation attaches to the song of the singer, that some of the feeling is projected, and that this is what gives the mate meaning. In our third case, the playful puppy bites his companion, and is sharply bitten in return. Pain enters into the coalescent situation as a whole. How is it distributed? In the phraseology of association, the nip he gives is closely linked with the pain he receives. By coalescence the pain and the nip form parts of the developed situation. But the companion is the centre of interest. And part of the pain is probably projected on this centre. That such projection actually occurs is rendered probable by such cases as the following, which was told me some years ago. A child, whose exact age I have forgotten if I then ascertained, was pricked by a pin, and he said, " Pin 'urted; poor pin." It is, indeed, not unlikely that with animals the outward projection of feeling is widely distributed over inanimate, as well

as animate, objects, and that its due restriction comes far later
in development, of which the so-called personification of life-
less things by savages may be a relic. In any case, the give-
and-take of play in young animals, and the after-earnest of
courtship and fighting, would seem to afford ample opportunity
for the external and internal distribution of feeling which
sows the seed in perceptual life of that which blossoms into
self and alter in the reflective life of ideational thought.

Although, therefore, an animal cannot conceive its companion as another self of similar nature, and with like passions
to his own, yet a considerable share of the feeling-element of
the conscious situation is projected on to that companion as
the chief centre of interest. And if it be said that this is
his feeling and not his neighbour's, the objection will be seen
to lose its force, so soon as it is realized that even man has
no experience of any feelings save his own. The only way we
can reach fellow-feeling is through sympathy ; and sympathy
has its roots in the projective process we have endeavoured to
describe. We endow our neighbours with natures as sensitive
to pain and pleasure as our own. This is a pre-requisite to
the social relationships termed ethical. But when we hear
people say, and find even Mr. Romanes putting on deliberate
record,* that " the feelings which prompt a cat to torture a
captured mouse can only be assigned to the category to which,
by common consent, they are ascribed—delight in torturing
for torture's sake," I venture to think that common consent,
if such it be, is wrong. As I said a dozen years ago,† before
Professor Groos had so carefully elaborated his theory of play,
" the cat or kitten plays with the mouse not from innate cruelty,
but for the sake of getting some little practice in the most
important business of cat life. Only man, who has the
capacity for nobler things, can be cruel for cruelty's sake ; "
and this is the direction in which Dr. Groos's opinion ‡ tends
to set. Mr. Romanes might have learnt a lesson in caution

* " Animal Intelligence," p. 413.
† *Atalanta*, Jan., 1889. Reprinted in " Animal Sketches," p. 17.
‡ " The Play of Animals," p. 122.

from his sister, who at first attributed a sense of shame to the
capuchin she so carefully studied, but subsequently was led
to adopt a simpler interpretation. " He bit me in several
places to-day," she says, in her admirable diary,* " but he
seemed ashamed of himself afterwards, hiding his face in his
arms, and sitting quiet for a time." She adds, however, in a
footnote : " On subsequent observation, I find this quietness
was not due to shame at having bit me ; for whether he
succeeds in biting any person or not, he always sits quiet
and dull-looking after a fit of passion, being, I think,
fatigued."

Shame is an ethical feeling. And as we have briefly discussed the germs of æsthetics in animals, so we may now as
briefly consider the germs of ethics. In its developed form
ethics is one of the " normative sciences " involving standards
of right and wrong. It is, as Professor Mackenzie says,† " the
science of the ideal in conduct." It involves a standard of
"ought," the product of reflection and generalization. Conduct
is compared with the ideal, and perceived to be either below,
up to, or perhaps beyond, the normal standard accepted by
civilized mankind. This involves a judgment ; and so far
as conduct is shaped in accordance with the ideal we attribute
the guidance to ethical motives. Such ideals, such judgments,
and the control of conduct through the play of such motives,
are probably beyond the mental capacities of animals. They
belong to the ideational stage of mental development, when
the conative tendency becomes volitional ; not to the perceptual stage, when it is impulsive. They do not enter into
the conscious situation as it takes form in the animal mind.
Behaviour has not in them acquired ethical meaning, since
in developed ethics, as normative, such meaning always has
reference to the norm, or standard. A real sense of shame
implies that our acts have fallen below our ideal.

It may be said that we cannot prove that animals do not
frame such ideals. But, if we accept the canon of interpretation

* Appendix to "Animal Intelligence," p. 486.
† " Manual of Ethics," p. 1.

above laid down, what has to be proved is that they do frame them. Is there any case among the hundreds that are popularly adduced to show that dogs are ashamed of themselves, that they possess a sense of justice, that they feel the prick of conscience, that on the one hand they know when they have done wrong, or on the other hand enjoy a sense of conscious rectitude—is there any particular case so described in the popular phraseology of anecdote, which could not be more simply described as the direct outcome of the coalescent situation, without the introduction of any implied reference to a standard of behaviour reached by reflective thought? The pug that has taken a nap on the drawing-room sofa, leaps down and slinks off with a "guilty" look on his master's approach. One can surely picture the previous situations, and be tolerably certain that they contained an element of reproof or something more energetic. The poodle that has successfully performed his tricks bounds to his mistress with an air of duty well performed. Has he never been petted and patted under such circumstance? Routine in many animals—so often creatures of habit—begets a customary sequence, the breach of which is at once felt. To this I ventured * to ascribe the conduct of the turnspit dog reported by Arago. He refused with bared teeth to enter out of his turn the drum by which the spit was rotated. The companion dog was put in for a few moments and then released ; whereupon the dog which before had been so refractory seemed satisfied that his turn for drudgery had come, and, entering the wheel of his own accord, began turning the spit as usual. The bared teeth may be here perhaps ascribed to an outraged sense of justice. But is it not a more simple, and just as probable, supposition that the behaviour was due to breach of customary routine. A trainer with whom I had some conversation on this matter pointed out a collie bitch, and said, "If I put her through her tricks in the usual order she does them like an angel ; but if I try and make her alter the order she snaps and sulks like the devil."

* "Animal Life and Intelligence," p. 404.

I have elsewhere[*] expressed my opinion that, though animals may behave in ways which may tend to mislead us, they do not act with intent to deceive. A dog is described[†] as "showing a deliberate design of deceiving" because he hobbled about the room as if lame and suffering from pain in his foot. But may not this be simply due to the fact that chance experience had led to a situation through which a hobbling gait had acquired the meaning of more petting and attention than usual? To behave with deceit as a deliberate motive implies the idea that the action will be interpreted as having a significance different from that which it really has. It is only possible on the ideational plane of mental development. It implies, too, from the ethical standpoint, a conscious departure from the standard of truth. The black that is acted has conscious reference and relation to the white that is not black. Few, however, will credit animals with deceit of this fully conscious and deliberate kind. Like the fibs of little children, the apparent deceit of animals is probably merely behaviour which has been associated in experience with pleasant results.

The case of shamming sickness, quoted from K. Russ, is thus interpreted by Professor Groos.[‡] And yet he adds, "When we see deception used so effectively to serve practical ends, examples of which are very common, it can hardly be doubted that there is in all probability more consciousness of shamming in play than we have any means of demonstrating." And elsewhere in the same work he observes,[§] "Many a grown animal still takes pleasure in the mock combats that he learned in youth. From a psychological point of view this phenomenon is especially noteworthy, from the fact that the adult animal, though already well acquainted with real fighting, still knows how to keep within the bounds of play, and must therefore be consciously playing a *rôle*, making

---

[*] "Animal Life and Intelligence," p. 400. "Introduction to Comparative Psychology," p. 369.

[†] "Animal Intelligence," p. 444.

[‡] "The Play of Animals," p. 299.   [§] *Op. cit.*, p. 145.

believe." I fail, however, to see the justification for the "therefore." Surely the difference of behaviour in this example, and in other such examples, is sufficiently explained as the outcome of diverse situations, without having recourse to anything so psychologically complex as the conscious self-illusion of make-believe—interesting and important as this is in the psychology of children. To suppose that a monkey who nurses a bit of blanket has any ideas about its being a make-believe baby is *not* to interpret the behaviour of animals in accordance with the canon we have adopted for our guidance.

To return to the "ethics" of animals. I have urged that ethical ideas, properly so called, have no place in their psychology. But just as the pleasure and satisfaction attending particular situations, as they severally arise, appear to contain the perceptual germs of what in later development becomes æsthetic appreciation ; so, too, do they also contain the perceptual germs of what becomes, through reflection in man, ethical approbation. And the situations in which these ethical germs must be sought are those which entail behaviour for the good of the social community. Indeed, we may go so far as to say that the perceptual foundations of ethics are laid in the social instincts. The satisfaction or dissatisfaction arising from the performance or non-performance of instinctive behaviour, evolved for the biological end of the preservation of the social community, is the perceptual embryo from which conscience is developed. Professor Mackenzie has indicated the ambiguities in the use of the term "conscience." "It is," he says,* "sometimes used to express the fundamental principles on which the moral judgment rests ; at other times it expresses the principles adopted by a particular individual ; at other times it means 'a particular kind of pleasure or pain felt in perceiving our own conformity or nonconformity to principle.'† This last seems to me," adds Mr. Mackenzie,

* " Manual of Ethics," pp. 285, 286.

† Starcke, *International Journal of Ethics*, vol. ii., no. 3 (April, 1892), p. 348.

"the most convenient acceptation of the term, except that
I should prefer to say simply that it is a feeling of pain
accompanying and resulting from our nonconformity to
principle." According to this definition the existence of a
principle or ideal is presupposed ; and the fact that Professor
Mackenzie lays stress upon the pain of nonconformity, shows
that the ideal is a high one. In the case of the animal,
however, such an ideal of right conduct has probably not
taken form. But Mr. Mackenzie also speaks of the "quasi-
conscience" begotten of custom. This comes nearer to the
feeling which animals may be supposed to have when their
behaviour does not accord with that which through instinct
or habit is the usage of the community. And if, as seems
to be shown by observation, animals sometimes punish the
breaches of such usage—when, for example, cats punish their
kittens for uncleanliness—the quasi-conscience will assume a
more developed form.

We may say, then, that the perceptual data are given in
animal experience from which, in ideational sublimation, ethical
ideals may be derived by a process of reflection and generalization. As in the case of æsthetics, so in that of ethics ; long ere,
in the course of mental evolution, the correlative conceptions
implied in the phrase "right or wrong" had taken definite
form, perceptual situations must have arisen in which
behaviour carried with it the feelings of satisfaction or the
reverse which laid the foundations of that approbation of the
right which forms the superstructure we build upon them by
the exercise of reflective thought.

V.—The Evolution of Feeling and Emotion

"Whatever conditions," says Dr. Stout,* "further and
favour conation in the attainment of its end, yield pleasure.
Whatever conditions obstruct conation in the attainment of

* "Manual of Psychology," p. 234. "Displeasure" here means the
feeling attitude antithetical to "pleasure."

its end, are sources of displeasure. This is the widest generalization which we can frame, from a purely psychological point of view, as regards the conditions of pleasure and displeasure respectively." Here Dr. Stout seems carefully to avoid the commonly accepted and much advertised conclusion, that pleasure and pain (to use this more familiar word as the antithesis of pleasure) are themselves the end of conative endeavour. And he is so far right that they by no means constitute the sole or indeed the primary end of all conative process. Attention is a conative act; but its primary end is not pleasure, but rather, as Dr. Stout says,* the fuller presentation of the object. No doubt this brings pleasure; but the fuller presentation comes first, and carries the pleasure with it. Instinctive response to felt stimulus falls within the conative attitude. In it there is that "inherent tendency to pass beyond itself and become something different," which Dr. Stout assigns to conation as its chief characteristic. But the end is not pleasure, but simply the instinctive behaviour. And if we say that the attainment of this end does bring satisfaction, which is a form of pleasure, Dr. Stout would probably reply that this is rather a result of the process than its true end.

Now, in such cases, what we are really dealing with is a class of *organic* processes having conscious accompaniments. No doubt the conscious accompaniments are of importance; they certainly cannot be neglected by the psychologist : but their feeling-tone does not constitute that which makes instinct run its course. And I have introduced the subject for present discussion in this way to reinforce what has already been repeatedly urged in the foregoing pages, that individual behaviour, in its first intent, is a biological legacy with ends predetermined through heredity. The inherent tendency to pass beyond itself and become something different, which for the old psychology was a heaven-sent impulse, or, as Addison said, "an immediate impression from the first Mover and the Divine energy acting in the creatures," becomes for the new

* *Op. cit.*, p. 65.

psychology an organic bequest. But the attainment of ends thus already predetermined has feeling-tone, both as process and in its resulting consciousness, and this feeling-tone serves to modify, through the situation it introduces, future behaviour, and thus, in a sense, affords a new end to subsequent conation.

"Life," wrote James Martineau,* "is a cluster of wants physical, intellectual, affectional, moral, each of which may have, and all of which may miss, the fitting object. Is the object withheld or lost? there is pain: is it restored or gained? there is pleasure: does it abide or remain constant? there is content. The two first are cases of disturbed equilibrium, and are so far dynamic that they will not rest till they reach the third, which is their posture of stability and their true end." This is an adequate description of the essential features in conative process. But in genetic precedence, as in individual development, the physical wants come first, and, at the outset of behaviour, the satisfaction or content is not and cannot be foreseen, since it has never yet entered into experience. To adopt a distinction suggested by Professor Mackenzie,† the conation is *purposive*, since we see that an end is involved, but not *purposeful*, since there is no definite consciousness of the end aimed at. But when experience has introduced feeling-tone into the situation, we may say that this, in a sense, introduces a new end to subsequent behaviour.

Mr. Herbert Spencer has said ‡ that pleasure is that which we seek to bring into consciousness and retain there; pain, that which we seek to get out of consciousness and keep out. May we assert, then, that, in the modification of behaviour due to experience, the pleasure to be gained or the pain to be avoided is the psychological end? Certainly not without qualification, unless we be among those who are content to accept any form of words which gives a general sort of notion

---

* "Types of Ethical Theory," vol. ii., f. 350.
† "Manual of Ethics," p. 85.
‡ "Principles of Psychology," vol. i., pt. ii., ch. ix. § 125.

of the kind of thing which we suppose is meant, and which is probably more or less correct. We want here and now to get clear ideas, and to express them with some approach to accuracy. To say that pleasure is the psychological end of intelligent behaviour is to put the matter too subjectively and in too abstract a form. Professor Mackenzie has clearly indicated the ambiguity in the word "pleasure." "Pleasure," he says,* " is sometimes understood to mean agreeable *feeling*, or the feeling of satisfaction, and sometimes it is understood to mean an *object* which gives satisfaction. The hearing of music is sometimes said to be a pleasure, but of course the hearing of music is not a feeling of satisfaction; it is an object that gives satisfaction. Generally, it may be observed that when we speak of 'pleasures' in the plural, or rather in the concrete, we mean objects that give satisfaction; whereas when we speak of 'pleasure' in the abstract, we more often mean the feeling of satisfaction which such objects bring with them." May we not go a step further, but entirely in the same direction, and say that pleasure is a constituent part of the concept self as an object of thought or desire; that its proper sphere is in the ideational consciousness ; and that, as we interpret the animal mind, it has no place as such therein ? The hedonist regards pleasure as the most excellent and distinctive characteristic of his ideal self and his ideal community. But animals have not risen or fallen to the level of hedonism. Pleasure is not for them a motive of conduct, though nice objects, as such, are attractive, and through them impulse acquires direction and force.

If, in animal psychology, we are to use the words pleasure and pain (as the antithesis of pleasure)—and they seem more properly to belong to a plane of mental development to which animals probably have not attained—we may say that the pleasure or the pain which attaches to any centre of interest in the situation is that which gives it attractive or repellent meaning; it furthers conation either towards or, as Hobbes would say, fromwards. But if we put the matter in

* "Manual of Ethics," p. 72.

this somewhat abstract form, let us keep in view, if it be only in the background of our thought, the kind of concrete example which may be adduced in its illustration—the dog with his attractive bone, the kitten that has raced off at sight of him, the cock-sparrow with trailing wings hopping after his mate, the falcon stooping on her quarry, the rabbit diving into his burrow at sight of the fox, and so forth. If we have such cases in view, where the centre of the situation has acquired or is acquiring meaning, a meaning which in large degree attaches to the external nucleus of the situation with only the germs of subjective reference, we may, perhaps, summarize the position by saying that in each case some pleasure to be gained or some pain to be avoided is the psychological end of conation.

But in each case the conation has also a biological end—the preservation and conservation of the race. "An animal," said Darwin,* "may be led to pursue that course which is most beneficial to the species by suffering, such as pain, hunger, thirst, or fear ; or by pleasure, as in eating and drinking, and in the propagation of the species ; or by both combined, as in the search for food." The important point here to notice is that the two ends agree—the psychological end of the attainment of pleasure and the avoidance of pain, and the biological end of race preservation. Under the joint influence of pleasure and pain, the needle of animal life sets towards the pole of beneficial action.

This consonance of end was in old days ascribed to the beneficent foresight of the Creator. The modern view, that it is a product of evolution, does not necessarily ascribe it to any other ultimate cause. For many still piously hold that evolution is only a name which we give to the method of creation. And there is not a fact or generalization in science by which such a conclusion can be disproved, for the premises lie outside the field of scientific inquiry. But the consonance of end is, for science, a remarkable fact, and one worthy of attentive consideration.

* "Life and Letters," vol. i., p. 310.

We have already seen that, if the claim for the inheritance of acquired characters be, on the evidence, judged unproven, and if instinct cannot be ascribed to transmitted habit, or regarded as a legacy of that which has been ancestrally acquired, the only scientific explanation of instinctive behaviour is one which involves the principle of natural selection. But no one doubts that, in the course of experience, animals acquire modes of procedure which are beneficial to the race. This is well seen in the play of animals as interpreted by Professor Groos. Now, why do animals play? From the psychological point of view, because they like it ; from the biological point of view, because they thus gain practice and preparation for the serious business of their after-life. But why do they like it? because, under natural selection, those who did not like it, and therefore did not play, proved unfit for life's struggle, and were eliminated. Suppose that an animal were born with a rooted hereditary aversion to everything nutritious and an inherited hunger for anything harmful and unfit for food. What chance would it stand of survival? Hereditary likes and dislikes determine the general course of acquired behaviour, just as hereditary nerve-connections determine the course of instinctive behaviour. Wherein, then, lies the difference between the two? In the fact that in the one case the nerve-connections are transmitted ready-made, while in the other they result from association or coalescence in the course of individual life. But in both cases the pursuit and attainment of the beneficial brings satisfaction.

Now, the consonance of end has long been regarded as an inevitable deduction from the hypothesis of evolution. "That pains are correlatives of actions injurious to the organism," wrote Mr. Herbert Spencer in his "Principles of Psychology," * " while pleasures are the correlatives of actions conducive to its welfare, is an induction not based on the vital functions only. It is an inevitable deduction from the hypothesis of evolution, that races of sentient creatures could have come into existence under no other conditions. Those races of beings only can

* Vol. i., pt. ii., ch. ix., § 124. I quote from the valuable "Epitome" prepared by Mr. Howard Collins, p. 214.

have survived in which, on the average, agreeable or desired feelings went along with activities conducive to the maintenance of life, while disagreeable and habitually avoided feelings went along with activities directly or indirectly destructive of life, and there must ever have been, other things being equal, the most numerous and long-continued survivals among races in which these adjustments of feelings to actions were the best, tending ever to perfect adjustment." And he safeguards the position by adding : " It is frequently taken for granted that the beneficial actions secured must be actions beneficial to the individual ; whereas the only necessity is that they shall be beneficial to the race."

This aspect of the consonance is now quite familiar ; but let us carefully note how completely dependent it is on natural selection. Mr. Herbert Spencer's testimony is especially valuable, since he has always laid much stress on the hereditary transmission of acquired characters and still holds * " that the inheritance of functionally-caused alterations has played a larger part than Darwin admitted even at the close of his life ; and that, coming more to the front as evolution has advanced, it has played the chief part in producing the highest types." Now, in these types we certainly find a wide range of consonance between the psychological and the biological ends of behaviour ; of which the phenomena of play may again be adduced as an example. Hence the special value of Mr. Herbert Spencer's testimony to the part played by natural selection in establishing the consonance. " Only those races of beings," he says, " *can have survived* in which, on the average, agreeable feelings went along with activities conducive to life ; " and again, " The most numerous *survivals* must ever have been among races in which these adjustments of feelings to actions were the best." The stress is here laid on the survival of those in which the consonance has obtained ; the elimination of those in which it was absent : that is to say, on natural selection. And where else can it be laid ? It is not the sort of thing which could be acquired. Suppose that, as we suggested above, an animal were

* " Principles of Biology," revised and enlarged edit. (1898), p. 560.

born with a rooted hereditary aversion to everything nutritious
and an inherited hunger for anything harmful and unfit for
food. Under what conceivable conditions could such an animal
acquire a complete change of its affective nature ? Animals
like things or they do not like them ; only to a very limited
extent, if at all, under natural conditions, can they learn to
like them. We, indeed, can in some degree learn to take
pleasure in that which at first, and by nature, is distasteful ;
but we do so by some external constraint, or from some motive
of ideational origin. We put pressure upon ourselves, or have
pressure put upon us, repeatedly to perform some irksome task ;
we fall into routine and custom ; and the performance becomes
so far second nature that its discontinuance produces an uncomfortable sense of something lacking in the daily round.
Perhaps domestic animals learn to like the good offices we force
them to perform for us. But here we have the element of
external constraint, which is wholly, or almost wholly, absent
under natural conditions. And there is no evidence that such
acquired likings are inherited. That, however, is another
question. Our present point is that, under nature, the conditions of such acquisition are lacking ; so that, there being no
acquisition, there is, in this case, nothing acquired to be
transmitted.

But, so far as behaviour is concerned, "functionally
caused alterations " are those due to the exercise of intelligence, by which the behaviour acquires direction and character
in reference to the meaning introduced into the situations.
See, then, the position to which we are logically driven. The
acquisition of that which has beneficial value in behaviour
depends on a consonance between psychological and biological
end. But this consonance is dependent on survival, and,
apart from special creation, or some kindred hypothesis such
as Leibnitzian harmony, can be due to nothing else. Even if
we grant, therefore, that the effects of acquisition are inherited,
the conditions of beneficial acquisition are dependent on
natural selection. And thus the inheritance of acquired
characters, which is so often urged as a principle of evolution

independent of natural selection, is, so far as intelligent behaviour is concerned, indirectly, if not directly, due to this very natural selection of which it is said to be independent. Surely, under these circumstances, the hypothesis in question may be said to be not only unproven, but altogether unnecessary.

And what is true of those diverse feelings which we group under the concepts pleasure and pain respectively, is true also of those more complex dispositions which we call emotional— using this term in a broad and comprehensive sense. We say that in their primary manifestations they are instinctive ; and they certainly seem to accompany organic behaviour due to co-ordinated reflex actions. But the emotion, as instinctive, is a matter only of its first occurrence. In the course of experience it enters into conscious situations, the centres of interest in which have acquired meaning.

Take a particular case.* Your dog is dozing on the lawn in the sunshine. Suddenly he raises his head, pricks his ears, scents the air, looks fixedly at a gap in the hedge, and utters a low growl. Place your hand on his shoulder, and you will find that his muscles are all a-tremble ; on his ribs, and you will feel how strongly his heart is beating. Soon the growing excitement leads to vigorous action, and he darts through the gap. You follow him across the lawn, look over the hedge, and see him facing his old enemy, the butcher's cur. They are moving slowly past each other, head down, teeth bared, back roughened. You whistle softly. Such a call would generally bring him bounding to your feet ; but now it is apparently unheard, at any rate unheeded. The two dogs have a short scuffle, and the cur slinks off. Your dog races after him ; and he flees, yelping. The situation is over. Spot returns, wagging his short tail, jumps up at you playfully, and then lies down again on the grass. But now and then, for ten minutes or so, he raises his head and growls softly.

Let us briefly analyze the dog's condition and actions, reading into them, conjecturally, the accompaniments in

* From " Animal Life and Intelligence," p. 382.

consciousness. As he lies on the lawn, he receives a sense-stimulus, auditory or olfactory. It has already acquired meaning, from many a tussle with the butcher's cur. It has organic effects, and it generates a conscious situation which has acquired complexity through coalescence. As the result of this situation the head is raised, the ears pricked, and so on. The dog is on the alert. His attention is aroused. The muscles of neck, eyes, ears, are brought into play in such a way as to bring the senses to bear on the exciting object. He probably sees the cur through the gap in the hedge. The muscles of the frame are innervated so as to be in a state of preparation to act rapidly and forcibly. At the same time the vaso-motor system is disturbed, the heart-beat is quickened, respiration is altered ; there is probably hardly an organ in the body which remains unaffected. Then the dog rushes through the hedge, and stands with bared teeth before his antagonist. A whole set of appropriate muscles are now strongly innervated. There is, perhaps, a double innervation, stimulating to activity and yet restraining from action. He bares his teeth and growls deeply. Attention is so concentrated that he heeds not, perhaps does not hear, his master's whistle. He is keenly on the alert. The blood-system, respiratory organs, and all his inner machinery are still pulsating with nervous thrills ; his back is up. Then he sees his chance, and flies at his opponent. Much that he has learnt in play, and all that he has learnt in earnest, comes to his aid in the short angry scuffle. And what we call his emotion of anger spurs him on to the fight; the cowardly dog in which this is lacking or is replaced by fear is spurred to flight. Each may contribute to self-preservation, but in different ways.

Now, we shall not attempt to determine how the distinctively emotional elements arise. Some think they arise by a sort of irradiating nervous diffusion in the nerve-centres as a direct result of the originating stimulus. Mr. Rutgers Marshall regards them as due to the motor activities in fight or flight ; Professor William James contends that they have their source in the visceral affections of heart, lungs, glands,

and so forth ; Professor Lange attributes them to vaso-motor
effects.   The problem is a difficult one, and hard to determine
by experiment ; for we have to deal with a matter of primary
genesis, of how they are at the outset introduced into the
conscious situation.   Experiments on animals which have
already gained emotional experience cannot decide the question of genesis.   Professor Sherrington, for example, has
shown * that, after severance of the spinal cord in the lower
region of the neck, and of the vagus nerves, by which "a
huge field of vascular, visceral, cutaneous, and motor reaction" were "deprived of all connection with the nervous
centre necessary to conscious response," "the emotional states
of anger, delight at being caressed, fear and disgust were
developed with, as far as could be seen, unlessened strength."
But the avenues of connection were closed *after* the motor and
visceral effects had played their parts in the *genesis* of the
emotion on the hypothesis that the emotion is thus generated.
Although new presentative data of this type were thus excluded,
their re-presentative after-effects in the situation were not excluded.   It is, moreover, an essential part of Professor James's
doctrine, as I provisionally accept it, that the "expression" and
the visceral and vascular efforts are independent results of stimulation in certain ways, and that these independent results are
conjoined through natural selection.   Suppose we sever the
connection through which the one takes effect, there is no
reason to expect that the manifestation of the other would
cease.   Professor Sherrington cut off the channels of communication with the visceral and vascular apparatus : if the
channels of expression remained open there is no reason why
such expression should cease.

We need not, however, for our present purpose, attempt to
ascertain how the distinctively emotional characteristics arise.
It is sufficient that they are presumably present in the situation.   Now, as Dr. Stout well points out,† the emotions

---

* "Experiments on the Value of Vascular and Visceral Factors for
the Genesis of Emotion," *Proc. Roy. Soc.*, vol. lxvi., pp. 390–403 (1900).

† "Manual of Psychology," p. 288.

generally presuppose the existence of certain specific tendencies. "The anger produced in a dog by taking away its
bone presupposes the specific appetite for food. The anger
produced in it by interfering with its young presupposes the
specific tendency to guard and tend its offspring. So the
presence of a rival who interferes with its wooing causes
anger because of the pre-existence of the sexual impulse." In
general, we may say that emotional states are, under natural
conditions, closely associated with behaviour of biological
value—with tendencies which are beneficial in self-preservation or race-preservation—with actions that promote survival,
and especially with the behaviour which clusters round the
pairing and parental instincts. The value of the emotions in
animals is that they are an indirect means of furthering survival. But how has the close association between emotional
condition and the biological end it furthers been established?
Again, we must say that under natural conditions it is not the
sort of thing which could be acquired. And again we must
urge that natural selection through survival is, apart from
some theory of pre-established harmony, the only hypothesis
in the field on which the close association can be explained.

There is one more point to which attention may be drawn.
If there be one thing, and there certainly are not many, on
which all writers on the emotions are agreed, it is as to their
vagueness. They do not readily submit to definition, and
cannot be described in a sentence. This is not due to any
indefiniteness of biological end, nor to much indefiniteness in
the mode of "expression;" it is due, rather, to an inherent
dimness and haziness of psychological outline. We seem
unable to focus them and get a clear-cut result. This is, no
doubt, in part due to the complexity of emotional states. But,
may it not be largely due to the fact that there is no necessity
for definiteness? They fulfil their purpose just as well if they
are vague. It is quite necessary for the dog to have a clear-
cut impression of his antagonist; and, on the cognitive side
of consciousness, meaning must be in some degree definite
to be of real value. But, so long as the emotion raises the

temperature, so to speak, to the boiling-point of vigorous action, it matters little what the psychological source of heat may be. If this be so, we should expect an emotional vagueness, since natural selection puts no premium upon emotional definiteness. And from this it follows, as a corollary, that, whereas we may infer that an animal's perceptual products are probably closely similar to our own, since sight, touch, hearing, smell, and taste are of value in so far as they convey definite meaning, in interpreting their feelings and emotions we have less secure grounds of inference, since all that is requisite is that there should be a sufficiently high emotional temperature to afford the conditions for definite and vigorous action.

In conclusion, then, we may say that the primary purpose of the evolution of feeling and emotion is to promote beneficial behaviour, and that the observed consonance of the psychological end of attaining satisfaction, and the biological end of securing survival, seems to be due to natural selection—is, indeed, scarcely explicable on any other naturalistic hypothesis.

A word of warning may be added. We have repeatedly spoken of biological and psychological ends. By this we mean what seems to the observer, as an interpreter of natural processes, the purpose and object of their existence. But the word " end " is often used in such a way as to imply foresight and contrivance on the part of a rational being. We have not used it in this sense. Whether the whole of nature, including animal behaviour, is driven onwards to definite ends by an underlying Cause, is a metaphysical question. It is not one on which science has any right to express an opinion one way or the other. Science deals with the phenomena ; the causes of their being lie outside her province.

CHAPTER VII

THE EVOLUTION OF ANIMAL BEHAVIOUR

I.—The Physiological Aspect

At the outset of our inquiry, we used the word " behaviour " in a wide and comprehensive sense. Thus broadly used, I said, the term in all cases indicates and draws attention to the reaction of that which we speak of as behaving in response to certain surrounding forces or circumstances which evoke the behaviour. The behaviour of living cells is dependent on changes in their environment ; that of deciduous trees, as they put forth their leaves in the spring or shed them in the autumn, is related to the change of the season ; instinctive, intelligent, and emotional behaviour are called forth in response to those circumstances which exercise a constraining influence at the moment of action. Used in this comprehensive sense, the term " behaviour " neither implies nor excludes the presence of consciousness. We know from our own experience, however, that consciousness does in some cases accompany behaviour, and we infer that in many other cases it may be present. But we need a criterion of its presence to guide our inferences, and this criterion we found in the ability of living beings to profit by experience. In Dr. Stout's phraseology, if a thing seems to acquire meaning for such a being, and the behaviour is guided in accordance with such acquired meaning, we infer the presence of consciousness as supplying conditions effective in determining its course. Still this does not exclude, nay, rather it presupposes, the presence of sentience at a lower stage of evolution, a sentience which is as yet ineffective since the process of conscious coalescence has not begun, or has not been carried far enough.

In foregoing chapters we have constantly held the problems of evolution in view, and in special sections directed attention to them. But the subject is so central to modern thought and discussion, that some further consideration of certain aspects of the evolutionary process and products will fitly serve to bring our inquiry to a conclusion.

We must accept, as a datum from the physiological point of view, the fact that protoplasm does respond to stimuli,—that it possesses the fundamental property of irritability. It is a substance that is in a state of unstable equilibrium. Its tendency to pass to a condition of more stable equilibrium is that in and through which organic behaviour in its very simplest expression is possible. And this, with progressive complication, runs through the whole gamut of animal behaviour, and eventually passes over into the sphere of consciousness. "The tendency to equilibrium," writes Dr. Stout,* "is the physiological correlate of what on the psychical side we call conation, —the striving aspect of consciousness." But, protoplasm at the outset—or as near the outset as we can get—is, in technical phrase, differentially responsive. The nature of the stimulus and the nature of the conditions decide what the nature of the response shall be. And even in that jelly-like speck of living matter, the *Amœba*, the responses conspire to a biological end. If they did not so conspire, we should not have the phenomena of life. The mere act of living, building up from food-stuff and oxygen an unstable substance which "explodes" and contracts under stimulation, implies that the processes which thus conspire are related in such a manner as to fulfil and secure their end. In higher unicellular animals, such as the *Paramecium*, the relations are less simple ; but in them the continuance of that sum of organic behaviour which we call life, is secured only on the condition that these less simple relations are duly preserved, and that the vital processes

* "Manual of Psychology," p. 132.

conspire with sufficient unity of biological purpose. And when we pass to the higher creatures in which many cells unite to form one animal, the very word "unite" indicates that the vital processes of all must conspire with sufficient *unity* of biological purpose to insure the continued life of the whole.

Now, in all the higher and more active animals a nervous system is developed, which has for its purpose and end the preservation and furtherance of unity amid circumstances of progressively increasing diversity. And in the course of its evolution an added means of preserving and furthering the essential unity is provided in consciousness, which, through the coalescence of scattered units of sentience, leads behaviour to acquire a new and higher unanimity of purpose. Thus a mental evolution is engrafted on the organic evolution which precedes it. But every step in this mental evolution presupposes a step in organic evolution. And such is the complexity of structure and process in all the higher animals that much of the business of behaviour is relegated to quasi-independent nervous centres, which perform this business automatically, and will continue to perform it, with much subsidiary unity of end, when they are left to themselves and all connection with the supremely unifying sensorium has been severed.

Before proceeding to give some examples of this fact, and to indicate its bearing on our interpretation of behaviour, it may be well to state distinctly that no attempt is or will be here made to trace in detail the course of the evolution of animal behaviour through the ascending grades of life, nor, indeed, to prove that there has been any such evolution. Evolution by natural genesis is here assumed as the only hypothesis with which science has any concern. If it be false, then have the labours of workers and thinkers, since Darwin and Mr. Herbert Spencer worked and thought, been vain. Special creation is not a scientific hypothesis, but a reference of biological and mental phenomena to an ultimate cause, which lies beyond and altogether apart from the scope of scientific inquiry. The fundamental assumption of the man of science is, that any natural event he may select for detailed study has natural conditions and antecedents. And it is only in such detailed study—taking this or that particular occurrence and endeavouring to ascertain what were its related antecedents—that advance in the evolutionary interpretation of nature can be secured.

Such advance has been secured by the labours of those physiologists who have established by careful experiment the quasi-independent action of subsidiary nerve-centres as constituents of the nervous system as a whole. In such animals as the crayfish and the lobster the central nervous system consists of a chain of "ganglia," or nerve-knots, which are connected together by nerve-strands. If these strands be cut between the thorax, which carries the walking limbs, and the abdomen or hinder portion of the body, the nerve-connection between these parts is severed. If the forepart be irritated through its sense-organs, the limbs of that part will respond ; but, whereas an unmutilated crayfish, subjected to such irritation, would give a vigorous flap of the tail, this does not take place in the crippled animal.* Still, if the abdomen be irritated, it will respond by a strong and swift contraction. The two portions of the body are each capable of acting independently with well co-ordinated movements, but no longer of working together with unity of purpose. In the hinder portion the abdominal limbs, or swimmerets, all swing backwards and forwards simultaneously with rhythmic strokes ; they act in concert. Sever now the connections between their ganglia, and each pair of limbs will continue to swing rhythmically but not with concerted rhythm. We have isolated a number of quasi-independent centres, and rendered them really independent. Each is concerned with its own proper co-ordination, but can no longer combine with others in a wider co-ordination. Mr. Hyde † has shown that in the king-crab,

* See Huxley's book on "The Crayfish," in the International Science Series, p. 108.

† *Journal of Morphology*, vol. ix. Quoted by Professor C. S. Sherrington in The Marshall Hall Address, "On the Spinal Animal" (reprinted from *Medico-Chirurgical Transactions*, vol. 82), p. 4.

Limulus, when the nerve-chain is severed just in front of the abdominal region, the rhythmic respiratory movements of the abdominal segments still proceed regularly and co-ordinately. Even when only a fraction of the nerve-cord, separated by severances in front and behind, is left, corresponding with a single abdominal segment, the rhythmic action of that segment continues ; but it is no longer synchronous with that of adjacent segments similarly isolated.

It will probably not be contended that the co-ordinated rhythm of the isolated segment in crayfish or king-crab is anything but a bit of organic and physiological behaviour. Whether it be accompanied by consciousness—a bit of consciousness isolated from other bits—we do not know ; but we have no grounds for supposing that the rhythmic behaviour is guided by consciousness.   And when, as Dr. Carpenter pointed out half a century ago, a water-beetle, from which the " brain " has been removed, swims forwards if placed in water, we must surely regard the co-ordinated progression as organic behaviour, whatever view we may hold with regard to a consciousness which in such a case is in a very literal sense a divided consciousness.

In these invertebrates the central nervous system is obviously segmented—one can distinguish the ganglia and their connecting nerve-strands.   In the vertebrate the brain and spinal cord form a continuous mass of nerve-tissue without obvious segmentation.   But the pairs of spinal nerves, each nerve with its afferent and efferent " root," indicate a really segmented condition, though in the cord itself the segments so run together and overlap that they cease to be externally obvious. And there is a certain, though limited, amount of overlap in the distribution of these segmental nerves.   Still, well co-ordinated responses occur when comparatively short portions of the spinal cord are isolated by severance from the rest.   In the male frog, especially during the breeding season, a clasping reflex is produced by stimulating the dark swollen pads on the inner side of the hand, and this, as Goltz has shown, is exhibited when all the central nervous system has been destroyed save the segments to which the nerves for the arms proceed. " Similarly," writes Dr. Sherrington,* " in the cat and monkey, the reflex wagging of the tail persists when behind the spinal transection only the sacral region of the cord is left intact." When the spinal cord of the dog is severed, so as to isolate that portion which is concerned with movements of the hinder part of the body, pressure on the pad of one hind-foot usually produces, not only a lifting of that leg, but also an extension of its fellow—that is to say, a co-ordinated response of the two limbs. But in the case of the vertebrates, more than in that of the invertebrates, the co-ordinated response of an isolated part of the central nervous system seems to lack the furtherance of its action, which normally comes from the higher centres from which it has been severed. "The spinal reflexes significant of progression seem," says Dr. Sherrington,† "to contribute chiefly towards preparatory posture in readiness for the onset of action executed by the musculature under the driving of higher centres. Thus the well-known reflex spinal posture of the frog is flexion of the hind limbs, the extensors of the joints being taut and ready for the jump. The spinal reflexes, which in their results approximate most closely to the normal reactions of the unmutilated individual, are those connected with the pelvic and abdominal viscera," many of which " are executed as spinal reflexes in a manner presenting little or no physiological defect from the normal. And if the bulb "—the continuation of the spinal cord within the skull to form the basal portion of the brain—" be included with the spinal cord, and these together, including their nerves, be isolated from the rest of the nervous system, the animal as regards its visceral life, including that of the heart and lungs, is practically intact."

Huxley graphically describes the actions of a frog from which the cerebral hemispheres have been removed. " If that operation," he says,‡ " is performed quickly and skilfully, the

* " The Spinal Animal," p. 5.

† *Op. cit.*, p. 23.

‡ " Collected Essays,' vol. i., essay on ' Animal Automatism," p. 224.

frog may be kept in a state of full bodily vigour for months, or it may be for years ; but it will sit unmoved. It sees nothing ; it hears nothing. It will starve sooner than feed itself, although food put into its mouth is swallowed. On irritation, it jumps or walks ; if thrown into water it swims. If it be put on the hand it sits there, crouched, perfectly quiet, and would sit there for ever. If the hand be inclined very gently and slowly, so that the frog would naturally tend to slip off, the creature's fore paws are shifted on to the edge of the hand, until he can just prevent himself from falling. If the turning of the hand be slowly continued, he mounts up with great care and deliberation, putting first one leg forward and then another, until he balances himself with perfect precision on the edge ; and if the turning of the hand is continued, he goes through the needful set of muscular operations, until he comes to be seated in security on the back of the hand. The doing of all this requires a delicacy of co-ordination and a precision of adjustment of the muscular apparatus of the body which are only comparable to those of a rope-dancer."

Now, why have we entered into these details ? To reinforce, from a somewhat different point of view, that which has again and again been urged in the preceding sections of this inquiry, that much of animal behaviour is an organic legacy. A going mechanism of great delicacy, with ready-made co-ordinations, the products of biological evolution, affords to consciousness a vast body of its primary data. As Dr. Sherrington himself says,* " co-ordination is abundantly shown to result from the independent power of the spinal arcs, altogether apart from the influence of the great cranial sense-organs, and of the cerebral arcs superposed upon them. These senses and the brain find the elementary co-ordination of the skeletal musculature an achievement already provided and to hand in the spinal cord. And no doubt the product of the instrument is, with the instrument itself, given over to their use in the reactions they elicit from the spinal musculature." We have seen how instinctive behaviour, in those animals in which it is

* *Op. cit.*, pp. 20, 21.

best studied, affords in hereditary biological outline, a sketch
which subsequent acquisition, under experience, serves only to
elaborate by the filling in of details and of the more delicate
shading in behaviour. But in the higher animals, in which a
period of youth is a time for the acquisition of experience—for
experimentation and practice,—it might seem that the inherited
biological legacy was of less importance. The "spinal animal,"
as Dr. Sherrington calls it,—that is, the animal in which the
spinal centres are isolated from the cerebral centres—goes far
to disprove any such view. In them the cerebral senses and
the brain find elementary co-ordination of the bodily movements an achievement already provided and to hand in the
spinal cord. But when different animals are compared—frog,
bird, rabbit, dog, and monkey—the permanent effects of severance of brain connection, the effects which remain when the
temporary period of disturbing "shock" is over, are more
marked in the higher than the lower types. And concerning
this, Dr. Sherrington says,* "The deeper depression of reaction
into which the higher animal, as contrasted with the lower,
sinks when made spinal, appears to me significant of this, that
in the higher types, more than in the lower, the great cerebral
senses actuate the motor organs, and impel the motions of the
individual."

Whether in the animal in which all direct connection between the anterior and posterior portions of the central nervous
system has been severed there is a double consciousness—a
cerebral consciousness and a spinal consciousness—we are not
in a position accurately to determine. If the generally accepted
opinion, that the higher brain-centres constitute the sensorium
or seat of consciousness, be correct, we must suppose that a
maimed consciousness, with many avenues of experience closed,
is retained in the anterior moiety, while the posterior is relegated to a condition of mere sentience at best. In any case all
relation between the two is prevented. The two—if two there
be—are rendered quite independent through severance of the
cord in the region of the neck. But there is a point of view

* *Op. cit.*, p. 29.

indicated by Dr. Sherrington which is full of suggestion and interest.

"It is significant in the evolution of animal form," he says,* "that the organ that exhibits most uninterrupted and harmonious increase in development, as studied successively in passing from lowest to highest, is the brain. And it is significant that in the nervous system—segmental system as it is— the brain is developed, not in those segments whose sense organs are ordinary cutaneous (tactual, etc.), muscular and visceral, but in the segments connected with the visual, olfactory, and auditory sense-organs ; in other words, the brain is developed in the head. The head is, so to say, the individual ; it has the mouth, it takes the food, including air and water, and it has the main sense organs providing data for both space and time. To this the body, an elongated motor organ with a share of the viscera and the skin, is appended primarily as a machine for locomotion. This latter must of necessity lie at the behest of the great sense organs of the head."

Now let us try and picture to ourselves a spinal animal, or one which retains only the lowest portion of the brain, the part known as the bulb or *medulla oblongata* ; let us assume that it is conscious and capable of acquiring experience through the association and coalescence of the data afforded by the senses that remain to it ; and let us try to imagine the conscious situations which would arise, and their value in the guidance of behaviour. The senses that remain are touch and the temperature sense, the motor sense affording data from the muscles, joints, and tendons, and those which supply certain visceral sensations. There is not one of much, if any, guiding value left. There is not one of what we may term anticipatory use. There is not one which could serve to infuse anything like definite meaning into the situation. For, after all, meaning is expectation. There is an element of anticipation in all those senses which are of any real guiding value in the conscious situation. Sight, hearing, and, especially for some animals, smell,—these are the senses which forewarn of

* *Op. cit.*, p. 18.

something which may follow; of other sensations with which in the course of experience they have coalesced. And they are all cut off from the supposed spinal animal. A light touch might in some cases forewarn of the shock or severe pressure, which would perhaps follow. But the shocks would so often come suddenly that it is questionable whether the warning would be of much avail. Still, touch is a warning and cognitive sense, and through it the environment would acquire a limited amount of meaning.

Now, the biological value of coalescent association lies in this very element of warning. The anticipatory senses, sight, hearing, smell, are in their several degrees the " projective " senses, the senses which carry with them the quantity of " outness." And their " projective " character is the necessary psychological expression of their distinctive biological end. They must be projective, must carry with them " outness," if they are to convey what we, following Dr. Stout, have so often spoken of as meaning. But if the biological value of coalescent association lies in the expectation it renders possible ; and if, in the spinal animal, there are no senses left save touch, which could receive from the environment preparatory warning of what is coming ; it would seem exceedingly improbable that it should develop quasi-independent conscious situations of its own. In the unmutilated animal, at any rate, tactual experience would most probably coalesce with that derived from the senses which more distinctively take the lead in the acquisition of meaning. And we may therefore, on these grounds, as well as on others, acquiesce in the current view that the quasi-independent functioning of the spinal cord and its constituent segments is, at best, lit up with those flashes of mere sentience of which Sir Michael Foster speaks in the passage we quoted in an earlier section.[*]

If from the consideration of the isolated spinal animal we turn for a moment to that of the isolated cerebral animal, we find it in a very different position. It is possessed of the warning senses—those which from several points of view are

<hr>

* *Vide supra*, p. 33.

the leading senses—but they are leaders without a following ; they have only a very limited company to conduct into action. The company is there, on the further side of the severance, but they have lost touch with it. They know not what it is doing, and have therefore neither the data nor the executive power to guide its manœuvres in the field of behaviour. They can form maimed coalescent situations, but they are as impotent as a mere theorist devoid of all power of practical application. We need not, however, follow the theme further. We need only add that, could we isolate tracts of nervous tissue in the lower brain-centres of such a cerebral animal, we should find that subsidiary co-ordinations would belong, as a physiological heritage, to these isolated fragments.

The conclusions we may draw, then, with regard to the evolution of behaviour, as viewed in its physiological aspect, are that it is, in its simplest expression and in its most complex, conditioned by sufficient unity of purpose to meet the biological end of survival ; that the complex unity of purpose may be analyzed into a multiplicity of subsidiary processes each with its subsidiary unity of purpose ; and that the psychological coalescence which gives unity to experience under the guidance of the leading senses, is paralleled in a physiological coalescence within the nexus of the nervous system.

II.—THE BIOLOGICAL ASPECT

The biological aspect of behaviour—its relation to biological ends—has so often come under our consideration in the foregoing chapters that little need be added in this section : and that little may be most appropriately devoted, first to the question whether consciousness does influence behaviour ; and secondly, this being accepted, to the importance of the *rôle* that is played by the development of conscious situations in securing, in the higher animals, the biological end of racial preservation.

That this end is secured without the aid of consciousness

in the case of many organic species, in all those, for example, which we classify as plants, must not be taken as presumptive evidence that in other species, for instance in the multitudinous host of insects, the development of conscious situations is of no biological value. The fact that chlorophyll is not developed in any mammal does not show that the possession of this substance is of no service to the higher plants. It would not be worth while to give expression to this very obvious truth, were it not that critics of natural selection persistently argue that because one species gets on perfectly well without this or that particular character it can have played no part in securing the survival of another species. When I described, at a meeting of naturalists, how well young chicks could swim, such a critic drew me aside after the meeting, and expressed his surprise that this did not convince me that the webbed foot of the duck could not logically be attributed to natural selection. This is an extreme case, and one obviously taken on peculiarly weak grounds. But even Huxley urged that, because a frog, from which the cerebral hemispheres have been removed, performs many co-ordinated actions without conscious guidance, consciousness is, throughout nature, merely an accompaniment of certain molecular changes in the brain. "Such a frog," he says,* "walks, hops, swims, and goes through his gymnastic performances quite as well without consciousness, and consequently without volition, as with it; and if a frog, in his natural state, possesses anything corresponding with what we call volition, there is no reason to think that it is anything but a concomitant of the molecular changes in the brain which form part of the series involved in the production of motion.

"The consciousness of brutes," he continues, "would appear to be related to the mechanism of their body simply as a collateral product of its working, and to be as completely without any power of modifying that working as the steam-whistle which accompanies the work of a locomotive engine is without influence on its machinery. Their volition, if they

* "Collected Essays," vol. i., p. 240.

have any, is an emotion indicative of physical changes, not a cause of such changes. It does not enter into the chain of causation of their actions at all."

If the literal truth of this contention—the logical soundness of this conclusion—be admitted, it seems absurd to speak of the biological value of consciousness in behaviour or to discuss the importance of the *rôle* that is played by the development of conscious situations in securing the biological end of racial preservation.

Now, consciousness is regarded by an influential school of thinkers as a sort of *deus ex machina*, which, sitting enthroned, and crowned with a capital letter as Will, directs, like a being from another sphere, the doings of the body. It was against the doctrines of this school that Huxley took up arms. They do not concern us here. The will, or volition, as an underlying cause, stands outside the pale of scientific inquiry. It belongs to the wide realm of metaphysics ; its plea must be heard in another court. In this part of his contention Huxley was, we believe, unquestionably right from the scientific standpoint. Neither will, nor impulse, nor instinct, nor consciousness itself, should be introduced into any scientific description or explanation of phenomena as a cause of their existence or being, for as such it does not enter into the sequence of events ; it is that which metaphysics claims as their *raison d'être* —that which gives them being. Science in this matter should be frankly agnostic—neither affirming nor denying aught. This, of course, is not equivalent to saying that the agnostic position is the true end of human reason. That would only be so on the assumption that the problems of science are the only problems with which that reason can deal. To exclude metaphysics from science is not to exclude it from human thought. As a matter of fact such exclusion is neither possible nor reasonable. But to clearly distinguish the problems of science from those of metaphysics is absolutely necessary, if we are to prevent hopeless confusion of issues.

In contending, however, against the introduction of metaphysical doctrines into the region of scientific explanation,

Huxley seems to have been carried too far by the force of his own attack. So long as he held to the position that every conscious state has, as its concomitant, a molecular change in the brain, he had all the forces of evolution on his side. But when he said that consciousness is merely the steam-whistle of life's locomotive, or merely answers to the sound which the animal-bell gives out when it is struck, he takes up another position of far less strategical strength. For whereas the frog from which the physical centres of consciousness have been removed sits crouched and motionless, and "will starve sooner than feed itself, although food put into its mouth is swallowed;" the frog in which conscious situations can take form in unmutilated cerebral hemispheres behaves in a very different manner. It is nothing less than pure assumption to say that the consciousness, which is admitted to be present, has practically no effect whatever upon the behaviour. And we must ask any evolutionist who accepts this conclusion, how he accounts on evolutionary grounds for the existence of a useless adjunct to neural processes.

"It is," says Huxley,* "experimentally demonstrable—any one who cares to run a pin into himself may perform a sufficient demonstration of the fact—that a mode of motion of the nervous system is the immediate antecedent of a state of consciousness. We have as much reason for regarding the mode of motion as the cause of the state of consciousness, as we have for regarding any event as the cause of another. How the one phenomenon causes the other we know as much, or as little, as in any other case of causation; but we have as much right to believe that the sensation is an effect of the molecular change, as we have to believe that motion is an effect of impact; and there is as much propriety in saying that the brain evolves sensation, as there is in saying that an iron rod, when hammered, evolves heat." But if we speak of the related antecedent as the cause, it is not obvious why we should not describe the desire to demonstrate the supposed fact as the cause of running in the pin. We seem to have just

* *Op. cit.*, pp. 238, 239.

as much reason for calling this antecedent state of consciousness
the cause of certain movements and behaviour, as of calling a
mode of motion in the brain the cause of a further state of
consciousness. It is true that we have not the least idea how
the desire can cause the act ; but Huxley practically admits
that we have no idea how molecular change can be the cause
of consciousness. In the one case we are no worse off than
we are in the other. Neither position is logically defensible ;
since each assumes that physical events and states of consciousness can constitute links in the same causal chain.

The philosophical hypothesis known as monism regards
the molecular change, not as the antecedent of a conscious
state, but as its concomitant. That which from a physical
and physiological point of view is a complex molecular disturbance is, at the same time, from a psychological point of
view, a state of consciousness. The two are different aspects
of one natural occurrence. Why such an occurrence should
have two so different aspects we have not the faintest idea ;
but here we are not one whit worse off than we were before.
The hypothesis does, however, help us to get over our difficulty.
An essential feature of Huxley's contention is that the physical
and physiological chain of causation is complete in itself,
which may be granted ; and further, that if consciousness does
arise it is merely an adjunct without influence on the sequence
of events—what is influential is the molecular disturbance, not
the consciousness which accompanies it. But according to
monism the state of consciousness actually is that very same
something which the physiologist calls, in the language of
physics, a molecular disturbance. And in saying that consciousness influences behaviour one who accepts this hypothesis
is merely avoiding a cumbrous form of circumlocution. He
puts it in this way instead of saying that the nerve-changes in
the cerebral hemispheres, or elsewhere, which from a psychological point of view *are* a conscious situation, influence and
determine the course of behaviour. But from this point of
view it is absurd to say that the consciousness is merely an
adjunct—absurd to say that were there no conscious situation

the neural situation would remain unchanged. They are the very same thing from different points of view ; and to say there is no influential conscious situation is simply equivalent to saying that there is not this determining neural situation.

However we explain the fact, there are few who hesitate to accept it for the purposes of scientific explanation. The conscious situation, having no doubt for the physiologist a neural aspect if he could only get at it as a whole, does practically determine the behaviour of the animal which has gained the requisite experience. If we accept the fact, we may pass on to its importance in securing the biological end of race preservation.

It is a commonplace of evolutionary doctrine that, other things being equal, those races will survive, in the constituent members of which intelligent behaviour enables them to deal most effectually with an environment of increasing complexity. And it is a matter of familiar observation that such behaviour is closely connected with delicacy and refinement of development in those senses which take the lead in cognitional process, and with rapidity and precision in the motor co-ordination through which prompt and skilful advantage is taken of the situation which has, through experience, acquired meaning.

But though the importance of intelligent adjustment to the circumstances of life is widely admitted as a general principle, it is perhaps through a study of animal behaviour that we are best able to realize its full range and extent. Biologists are so largely, and quite wisely, occupied in the study of morphological and physiological problems, which admit of a treatment more exact than the most ardent advocate of the investigation of behaviour, under natural or even under experimental conditions, can claim ; they devote, again quite rightly, so large a share of attention to the variation and natural selection of adaptive structure in its adult condition and embryonic stages ; the pendulum of opinion has, under the teaching of Professor Weismann, swung so far in the direction of the non-acceptance of the hereditary transmission of characters individually acquired through intelligent adjustment or otherwise ; that the part played by consciousness in the evolution of the higher and more active animals is apt to pass unnoticed or unrecorded. It is well, therefore, to put in a reminder that a great number of animals would never reach the adult state in which they pass into the hands of the comparative anatomist save for the acquisition of experience, and the effective use of the consciousness to which they are heirs ; that their survival is due, not only to their possession of certain structures and organs, but, every whit as much, to the practical use to which these possessions are put in the give and take of active life ; and that many interesting problems which are keenly discussed by evolutionists in the light of natural selection presuppose conscious situations which are more or less tacitly taken for granted.

Let us cast a rapid glance over some of these topics of biological discussion. The fascinating subject of mimicry, involving as it necessarily does the discussion of the value of warning colours and behaviour, a subject opening up an extensive group of problems so brilliantly studied by Professor Poulton, is meaningless save in so far as there is implied a conscious reaction to colour and form on the part of animals which can learn from experience. The warning colours reinstate a conscious situation, so that, misled by appearances, a bird mistakes the mimicking insect for its nauseous " model."

The whole range of behaviour, included under play, experimentation, and practice, on the importance of which, following the lead so ably given by Professor Groos, we have insisted, is equally meaningless, save as a means to the acquisition of serviceable experience for use in the more serious business of after-life ; and experience is the establishment, through association and coalescence, of conscious situations which possess guiding value. And if, as we shall hereafter see, they may also be regarded as a means of securing pleasure, as a psychological end of behaviour, it is not less obvious that it is only through the development of consciousness that such a psychological end can have any existence.

It matters not if the particular form assumed by play and experimentation be largely dependent on instinctive tendencies. For all the phenomena of instinct, profoundly organic as are the modes of behaviour comprised under this head, definite as are the inherited co-ordinations in the most typical examples of its occurrence, have also, except in some doubtful cases, a conscious aspect. At any rate this is the case in so far as instinctive response forms the hereditary basis on which is reared a more nicely adjusted intelligent edifice, in so far as instinctive procedure is subsequently modified and guided by acquired experience, in so far as there creeps in that "little dose of judgment" which Huber found in bees, Lord Avebury attributes to ants, Dr. Peckham sees in spiders and solitary wasps, and all observers find in birds and mammals. For if in these cases instinctive behaviour were unconscious, it would, as such, remain outside experience ; and if outside experience, there could be no data on which consciousness could base any modification of inherited behaviour, no opportunity of taking up the ready-formed responses into the mental synthesis and utilizing them for the wider ends of intelligent purpose.

In social behaviour there is a reciprocity of suggestion between the members of the community. And such suggestion is operative through an appeal to consciousness. However instinctive the forms of procedure may be in social insects, there remains much beyond which is hard to explain on the hypothesis that there is, in them, nothing analogous to a conscious situation ; while in such vertebrates as birds and mammals we cannot but believe that consciousness is the main determinant of much behaviour which seems to imply the germs, or more than the germs, of sympathy. The little monkey I saw in Hamburg cuddling up caressingly to a wounded companion, must surely have experienced a conscious situation analogous to that which prompts a child to nestle alongside her companion in distress. And he who has seen no signs of sympathy in dogs, has either watched their behaviour in vain, or is himself lacking in sympathy.

In sexual selection by preferential mating, even if we follow Professor Groos in believing that it is a special mode of natural selection, the conscious situation is essential. If we accept the theory in any form, we must regard the adornments, antics, and display of the male as an appeal in some way to the consciousness of the female, whatever particular form the effects in that consciousness may take, whether the appeal evoke a sense of beauty, or simply be a means of exciting to the consummation of the natural end of courtship. Even if we follow Mr. Wallace in regarding plume and song as " recognition marks," it is only by their appeal to consciousness in this way, if in no other, that they are of any biological value. And this, of course, applies equally to the whole range of his theory of recognition marks—their sole utility lies in their being a stimulus to consciousness through which the end of recognition is secured. So, too, not only the specialized behaviour which we dignify by the name of " courtship," but every case in which mate is drawn to mate through sight, smell, hearing—any of the leading senses—testifies to the importance of consciousness in furthering an end of supreme biological importance.

And if, as Darwin urged, the " law of battle " among the males co-operates with preferential mating, as we can hardly deny, in securing strong, vigorous, and healthy fathers of the generation they beget, here, too, consciousness is an important factor. Can we conceive a " law of battle " among unconscious beings ? If success in the combat were a mere matter of brute strength, it would imply some consciousness in its dull exercise. But it is more. It is also a trial of skill. Were it not so our forefathers would not have spent hours in watching a cock-fight, or laid heavy odds on their particular " fancy."

We need not labour the theme. In the search for food or a nesting site, in the capture of prey and escape from enemies, in all that demands attention, and in all that necessitates practice, in what M. Houssay calls " the industries of animals," and in that which Mr. Hudson calls " tradition," consciousness has a part to play. Even plants unconsciously appeal to the consciousness of insects, birds, and mammals. Their bright, scented, nectar-bearing flowers, and their sweet, coloured fruits are means of effecting the biological ends of fertilization and the dissemination of seeds, but only on condition that their colours stimulate the sense of sight, and their scent and sweetness the senses of smell and taste. It is, perhaps, going too far to claim that, wherever sense-organs exist they imply at least some dim and rudimentary form of conscious situation of guiding value so far as it goes ; for it is possible that in some cases the coalescence of elementary items of sentience has not been carried far enough to justify us in speaking of experience by which the animal can profit. But it is surely not going too far to claim that, wherever two or three such sense-organs are gathered together in any living being, there is consciousness in the midst of them, beginning to exercise that guidance which serves so markedly to differentiate the typical animal from the typical plant.

But throughout the animal kingdom, until we reach its highest development in man, the guidance of consciousness, important as it is, seems to be almost wholly subservient to a biological end, that of the preservation of the race, and for the race of the individual. Practical utility is the touchstone of animal intelligence, and of the whole range of feeling and emotion in beings still under examination in the stern school of natural selection. By this we mean that practical utility has determined what degree and complexity of intelligence, feeling, and emotion shall be attained. If the requisite level be not attained—elimination. Higher levels no doubt bring advantage—so long as they are practically useful. But in the school of natural selection useless accomplishments are not much taught. Although its examinations are in a sense competitive, all are allowed to pass who qualify for survival. But the competitors become more numerous and the standard for a pass rises. As the school increases in size higher classes with harder problems to solve are established. Progress is an incident of the constant survival of the fittest when there are variations in fitness.

III.—THE PSYCHOLOGICAL ASPECT

On the hypothesis of monism, the nature of which, so far as it bears on our inquiry, was briefly indicated in the foregoing section, the conscious situation is the psychical or mental expression of that which for the physiologist is what we may term a neural situation. As such it does not enter into the chain of physical causation ; nor do physical events as such—that is to say, save as experienced—enter into the chain of mental causation. For mental development they have no independent existence, and are negligible except in so far as they enter as items of experience into the conscious situation.

But altogether apart from the way or ways in which we may attempt to explain the fact, most of us believe, with unquestioning confidence, that the growth of practical experience, somehow associated with nervous changes in the brain or sensorium, is of real value in the guidance of behaviour in such manner as to secure biological ends. Conscious experience must therefore, in the animal world, serve its biological purpose, or it will be of no avail. If there be not a pre-established harmony, there must be an evolved harmony ; and how such a harmony could be evolved if consciousness be not by some means in vital touch with behaviour, influenced by and in turn influencing it, we cannot conceive. The steam-whistle theory of consciousness leaves the matter, for the evolutionist, in this inconceivable position.

We need not, however, flog a dead horse. We need not ask how, on the steam-whistle theory, those states of feeling which we broadly classify as pleasurable could become associated with behaviour conducing to welfare, and those which we group as hurtful with behaviour which is biologically harmful. It is more important, again, to notice that, associated and consonant with the biological end, there arises a psychological end of behaviour—what we may term, with the qualifications before considered,* the getting of pleasure and the

* *Vide supra*, p. 285.

avoidance of pain. This is the purpose of behaviour as viewed from the psychological aspect. The biological end of animal conation is racial survival ; its psychological end is individual satisfaction. And the two ends are, in the main and broadly speaking, consonant—a result which would unquestionably be secured by natural selection, but is on any other naturalistic hypothesis difficult of explanation.

But the two ends are not only consonant ; they are supplementary one to the other. During much of the life of the higher animals there is no need, immediately present and pressing, for the output of action to meet biological ends. There are periods of life and intervals of time when the sharp incidence of the struggle for existence does not call for the serious business of behaviour. But at these periods and in these intervals the animal is not inactive ; indeed, it is restless in its activity. Unless it be weary with unwonted exertion, or basking in the psychical sunshine of content, due to the unsought advent of pleasant stimulation or the after-effects of previous behaviour (for example, when hunger has been relieved), the healthy animal must be up and doing. This familiar fact no doubt affords the basis in observation of the surplus-energy theory of play. But is it necessarily surplus energy ? Is it not rather normal energy which expends itself in this way when there is no immediate and serious biological business on hand ? And, as Professor Groos has pointed out, play is seen when we have every reason to suppose there is no surplus energy, nay, even when the normal energy is at a low ebb. There is no more pathetic sight than a sick kitten, with energy obviously much below par, utilizing its little remaining strength in feeble attempts to play.

It is unnecessary to do more than remind the reader of the theory elaborated with so much skill and care by Professor Groos, that the forms assumed by play—in which, it will be remembered, he includes a very wide range of behaviour—have a very important indirect biological end in practice and experimentation. Our present point is, that its direct psychological end is the satisfaction it affords. Without this the individual would not be impelled to the continuance of performances which occupy a wide space in the field of animal behaviour in which the biological end has reference, not to present requirements, but to future needs.

No one has given better expression to the sway of this psychological end than Mr. W. H. Hudson. "We see," he says,* "that the inferior animals, when the conditions of life are favourable, are subject to periodical fits of gladness, affecting them powerfully and standing out in vivid contrast to their ordinary temper. And we know what this feeling is—this periodic intense elation which even civilized man occasionally experiences when in perfect health, more especially when young. There are moments when he is mad with joy, when he cannot keep still, when his impulse is to sing and shout aloud and laugh at nothing, to run and leap and exert himself in some extravagant way. Among the heavier mammalians the feeling is manifested in loud noises, bellowings, and screamings, and in lumbering, uncouth motions—throwing up the heels, pretended panics, and ponderous mock battles.

"In smaller and livelier animals, with greater celerity and certitude in their motions, the feeling shows itself in more regular and often in more complex ways. Thus Felidæ, when young, and in very agile sprightly species, like the puma, throughout life, simulate all the actions of an animal hunting its prey—sudden, intense excitement of discovery, concealment, gradual advance, masked by intervening objects, with intervals of watching, when they crouch motionless, the eyes flashing and tail waved from side to side ; finally, the rush and spring, when the playfellow is captured, rolled over on his back, and worried to imaginary death. Other species of the most diverse kinds, in which voice is greatly developed, join in noisy concerts and choruses ; many of the cats may be mentioned, also dogs and foxes, capybaras and other loquacious rodents ; and in the howling monkeys this kind of performance rises to the sublime uproar of the tropical forest at eventide.

* "Naturalist in La Plata," pp. 280, 281.

"Birds are more subject to this universal joyous instinct than mammals, and there are times when some species are constantly overflowing with it ; and as they are so much freer than mammals, more buoyant and graceful in action, more loquacious, and have voices so much finer, their gladness shows itself in a greater variety of ways, with more regular and beautiful motions, and with melody. But every species or group of species has its own inherited form or style of performance ; and however rude and irregular this may be, as in the case of the pretended stampedes and fights of wild cattle, that is the form in which the feeling will always be expressed."

That all this, which Mr. Hudson so graphically describes, belongs to the psychological aspect of animal behaviour and is directly prompted by conative tendencies whose immediate end is conscious satisfaction, the mere joy of unrestrained and healthy activity, may be freely admitted, without denying that all this exuberant psychical life owes its evolution to the fact that it is in consonance with and supplemental to biological ends which secure survival. It is with animals as it is with man ; play is the preparation for earnest. As I have else-where said,* what our national games have done for the English race it is difficult to overestimate. They train us to use our bodies and expend our energies to the best advantage. An old soldier, watching a football match, said, "That's the training for our future soldiers and sailors." The playing fields are the finest school of organized co-operation in the world. But, apart from compulsion, a boy will not enter into the game with that zest through which alone it acquires real value for training, unless there be an immediate psychological end in the satisfaction he derives. And with animals practice and preparation for the business of life could not occur if the ultimate biological purpose of it all were not supplemented by the enjoyment it brings for its own sake.

But in animal play, as indeed in that of human youth, we are perhaps a little apt, in laying stress on the bodily skill and

* "Psychology for Teachers," p. 70.

readiness of response to which it so effectually ministers, to forget that it is also a psychological training. In technical phraseology, we are disposed to fix our attention on the acquired co-ordination of act and movement rather than on the correlation of conscious data, which renders possible the skilful performance. And yet, rightly considered, the behaviour itself is simply the outcome of a conscious situation, duly elaborated, and knit together through the association and coalescence of its constituent data. It is a means to the unification of consciousness by bringing into relation scattered and, at first, quasi-independent sensory and emotional elements. Success is only attained through the concentration of attention and effort on that which is the centre of interest and also the focus of endeavour. And this close attention and well-directed effort, which are trained in the playful output of energy, are just the mental qualities which will stand the animal in good stead when the real incidence of life's struggle comes upon it, when the reward of success is survival and the penalty of failure elimination. For they are not merely physical qualities, though their effects are bodily movements of attack and defence, of active escape, or merely " lying low." They are essential psychological features of a unified and well-directed conative process.

In the fairly abundant play-time of animal life, this unification and direction of conative process can take form under conditions wherein the preliminary failures which accompany all forms of learning do not entail the severe penalty of elimination. If we may so put it, and so apply a deeply instructive parable, Natural Selection says to her more favoured children, in which conscious situations can be developed, " Here are the talents with which I have endowed you ; make use of them till I come, as come I shall in due time." This animal puts them out to usury in play ; that animal keeps them laid up in the napkin of inactivity. Then Natural Selection, the austere one, comes ; gives the commendation of survival to the animal that had learnt to put its talents to use in the period of preparation, and condemns to elimination that which had not traded with his talents at the bank of play. In animal life, on the perceptual plane, we have the same need for training in little things and seemingly unimportant matters in preparation for the stress and storm which may, nay must, come upon them, that we find in men and women on the higher ethical plane. To those who think that the play of animals is too trifling a thing to affect the question of survival, we would suggest the application, with a necessary difference, of the thought which Miss Edith Simcox puts into the following words : " Does it," she says, " seem a trifling thing to say that in the hours of passionate trial and temptation a man can have no better help than his own past ? Every generous feeling that has not been crushed, every wholesome impulse that has been followed, every just perception, every habit of unselfish action, will be present in the background to guide or to sustain. It is too late, when the storm has burst, to provide our craft with rigging fit to weather it ; but we may find a purpose for the years that oppress us by their dull calm, if we elect to spend them in laying up stores of strength and wisdom and emotional prejudices of a goodly human kind, whereby, if need arises, we may be able to resist hereafter the gusts of passion that might else bear us out of the straightforward course." To apply the thought, the trifles of play supply the psychological rigging which alone can save the animal craft in the coming storm of the struggle for existence. And the point on which we have to lay special stress in this section is, that it is psychological rigging—or, if this seems to lay too much emphasis on the genesis of conscious situations, we may at least urge that the psychological ropes are of co-ordinate importance with the biological spars.

So far, then, we reach the following conclusion : that if we classify the behaviour of the higher and more intelligent animals under two heads, the one comprising all those acts which are of direct biological value in enabling the animal to escape elimination under the immediate stress of the struggle for existence, and the other including all those acts which are of indirect preparatory or educative value, the latter, which are under their biological aspect not less important than the former, are under their psychological aspect of perhaps even greater importance. For the conditions of actual struggle are not those under which mental development could most easily be furthered, though they are those in which it is most effectually tested. Hence, the more intelligent animals pass through a period during which they are more or less shielded from the incidence of natural selection by their parents, and this is the period of play and of psychological education. And the tendency to play is so far organic, in that it is dependent on inherited instinctive propensities, and so far psychological in that it is accompanied by a felt want, which constitutes a conative impulse finding its appropriate end in the consciousness of satisfaction. But play—if we accept the term as the group-name for all those modes of behaviour which fall under our second class, those of indirect biological value—does not cease with the period of youth ; it occupies all the intervals in the more serious business of animal life. And no discussion of animal behaviour can be adequate which does not assign to this class its due place, alike in biological and in psychological evolution.

The whole value of experience lies in the linkage and coalescence of the data afforded to consciousness. It is true that an inherited nervous system supplies the organic conditions of that physiological linkage and functional coalescence of which experience is the psychological expression. It is true that this physical integration secures a ready-made grouping of the conscious data which are the concomitants of orderly molecular changes in the brain or analogous sensorium. Still, it also remains true that the value of experience lies in the further linkage and coalescence that is acquired by the individual in the course of what we may fitly call its education. Every step in this education gets its psychological sanction through the satisfaction it affords in consciousness ; and the time of acquisition is not during the stress of examination in the actual struggle for existence, but rather in the youthful period and in the subsequent intervals of preparation and practice, during the play-time of animal life.

The examination analogy—if, indeed, it may not be rightly regarded as something more than an analogy—may be pressed a little further as a means of fixing our attention on two points which are worthy of consideration. The first is that, in the preparation for the examination, specific practice as much of it is, cramming is not the system exemplified by the higher animals. A good all-round education in the acquisition of conscious situations more or less coalescent into a unified system of experience, and in their effective utilization without unnecessary delay and bungling along more or less converging lines of practical behaviour; this is what secures a " pass " in survival, especially where the circumstances of life have reached a considerable degree of complexity. The instinctive act, with its relatively definite response to a question which is almost certain to be set to every candidate for survival, is that which is the analogue in behaviour to the result of a system of cram. Organic nature does employ this system in the lower classes of her school; definite responses are ground into merely instinctive types generation after generation, and the right answers are given, automatically and unintelligently, whenever the oft-recurrent questions are set. But this will not do when the questions require the exercise of intelligence, when they are of the nature of problems, with just those delicate but not unimportant shades of difference which baffle the candidate who has been drilled in a merely mechanical fashion. Hence the cramming of instinct does not suffice for animals whose environment presents problems of greater variety and greater complexity. Intelligence is required to meet the particular combinations as they arise. The greyhound, which is loosed on a hare, has never seen that hare run in exactly that way over that special tract of country. But he has been trained in such situations, and is thus prepared to meet the special problem in its details as they present themselves in the light of the experience he has gained of other like problems. And his skill in pursuit has not only been gained through education in coursing. In a thousand ways, as puppy and dog, he has learnt how to use well those sinewy limbs. The training of his whole life is brought to bear on the question immediately before him.

The general bearing of these facts is obvious. Play, as a means of animal education, is varied, and has for its end all-round training of the animal mind in its sphere of operation. Although there are some specific propensities, certain observable trends of behaviour, as in hunting-play, courtship-play, and the like, we must not expect, nor do we find, anything like stereotyped definiteness of conative activity. We find that freedom and elasticity in animal education which is, perhaps, more often advocated than carried into practice in human education.

The second point arising out of the examination analogy is, that its range determines the level of preparation therefor. It is, for animals, a practical examination, not a theoretical. Not a single question is set demanding an explanation. The problems are such as can be solved by intelligence, not such as require the exercise of reason, as we have used the term in foregoing pages. These higher problems are only set when the sixth form is reached, and there is no conclusive evidence that any animals get into the sixth. This, however, is entirely a question of evidence, and many of us will be glad to welcome them there, if proved ability to deal reflectively with ideational questions justifies their promotion.

If any of them do belong to this form, they have probably got there through play. For in the stress of the actual examination there is not much time for reflection. Or perhaps we may rather say that, not in actual struggle, and not in active preparation for it in play-time, but in intervals of leisure between both, when the animal lies quietly turning over in his mind we know not what, will experience be reviewed, and generalizations drawn as to the why of events in this strange world. Probably the animal accepts things as they are, and does not trouble about their explanation. But it may not be so. At any rate, if animals lack the means of descriptive inter-communication, and have no words as concrete pegs on which to hang abstract ideas, their explanations cannot be carried far. Theories without the power of disputation would be a poor solace in leisure moments.

One more point may be noticed with regard to the psychological aspect of the evolution of behaviour—the reciprocal action of intelligence. It is the intelligence of others that introduces so much variety and complexity into the environment. Hunters and hunted, combatants, rivals, mate and mate, enemies or companions in their varied aspects, introduce through their intelligence complications which only intelligence can meet. And, as intelligence begets intelligence, so do emotional attitudes beget answering emotional states. Psychological evolution translated into practical behaviour gives rise to situations of reciprocal complexity. This point of view is, however, so familiar, that nothing need be said in its further elucidation. The behaviour of any given animal does not stand alone, but is closely related with the behaviour of others. Among social animals the relationships are peculiarly close, and it is among them that the psychological aspect of behaviour reaches its highest expression.

IV.—Continuity in Evolution

Under the head of organic behaviour, in the widest acceptation of the term, fall the whole of physiology, the whole of embryological development, nay, more, the whole of organic evolution ; while mental evolution, in all its stages, may be regarded as the psychological aspect of that which, from the physiological aspect, is the evolution of nervous systems. Life itself is the behaviour of a particular kind of substance which is found more or less abundantly under natural conditions. No other known substance behaves in this way, and so ignorant are we as to the conditions of its natural origin, that it is useless to guess at a scientific explanation. And even if we knew all the antecedents and conditions of its origin we should be no nearer a comprehension of why protoplasm has the peculiar properties which we find it to possess. That is a question to which science can give no answer. Who knows why a certain compound of oxygen and hydrogen in certain proportions has the properties of that which we call water?

Let us note the distinction between saying, as we said above, that life is the behaviour of protoplasm, and asserting that life is the cause of this behaviour. The one is a scientific statement of observed fact, the other an explanation of the fact in metaphysical terms, a reference of the fact to its underlying cause. So long as we quite clearly understand that we are talking the language of metaphysics, we may speak of life as a cause of organic behaviour ; but let us be careful to remember that the statement has no more value for science than the assertion that aqueosity is the cause of the behaviour of water.

Leaving on one side, then, the natural origin of protoplasm, the conditions of which are unknown, we find that, as a matter of observation, every bit of living substance, the history of which has been traced, is a fragment detached from some other bit which behaved in the same way. This is the basal fact of the continuity of organic evolution. But such a detached fragment has the property of increasing by taking up from the environment more of those elementary materials from which it is itself compounded in subtle synthesis. Nay, further, every fragment of which we know the history is found to increase in such a way as to reach, in form, structure, and idiosyncracies of behaviour, the likeness of the organism— plant or animal—from which it was derived. In the higher plants and animals the separated fragments or cells are the ova and sperms, or their equivalents, which unite, with fusion or coalescence of their nuclear matter, and thus give rise to a new individual in the course of embryological development.

Now, as we have already seen, much modern biological discussion centres round the question whether the detached reproductive fragment, ovum or sperm as the case may be, is derived from the whole body of the parent, by what Darwin termed pangenesis or in some other way, or only from germinal substance set apart in development for this end. And we have provisionally accepted the hypothesis that it is the direct descendant of other reproductive cells ; and that, throughout a long ancestry, stretching back into the far past, there never occurs in the direct line of genealogical sequence, any highly differentiated. cell, such as a gland-cell, muscle-cell, nerve-cell ; never, with certain reservations into which we need not enter, is found the representative of any tissue save that to which the reproductive function is restricted. In technical phraseology, the continuity of organic evolution is due to the continuity of the germinal substance.

During embryological development the fertilized ovum—consisting of two fused fragments of this germinal substance—gives rise to a host of ordered and marshalled cells, which are divisible into two groups : the one forms the body with its muscles, bones, glands, digestive system, skin, sense-organs, nerve-centres, and so forth ; the other forms.a reserve store of germinal substance, from which are derived the ova and sperms. The former take no direct share in reproduction ; they are off the line of continuous descent ; they die without issue. But they protect and minister to the reproductive function of the second group—the potential ancestors of the races to follow. But all instinctive and intelligent behaviour is the outcome of the orderly working of the nervous system, is initiated through sensory stimulation, and is executed by the motor organs ; and all the structural parts, through which such behaviour is possible, belong to the body—that which dies without issue. How, then, can instinct and intelligence be inherited ? In a sense they are not inherited. The nervous system which is their organic basis begets no heirs. But it is begotten of germinal substance, which not only produced the body of which the nervous system is a part, but also handed on, with that body, samples of the same germinal substance capable of reproducing a similar body and a like nervous system. Herein lies the basis of heredity.

The stress of the struggle for existence falls upon the body ; and instinctive or intelligent behaviour is a means to its preservation in the struggle for existence. According as it survives or not, will the samples of germinal substance it

contains fulfil their biological end or perish with it. Natural
selection secures the survival of those animals which bear the
seed from which their like will be developed.

On this view all variation arises within the germinal substance, but it is manifested in the body which is its product.
How variations arise we do not know with any exactness of
detail. That the germinal substance is influenced in its nutrition and in other ways by the surrounding tissues is highly
probable ; and this influence may lead to changes which are
the source of variations ; but it is very doubtful whether such
influence can be what we before termed " homœopathic." * It
is improbable that the formation of the nerve-connections
involved in intelligent behaviour which has grown habitual
through repetition, can so influence the germinal cells as to
give rise to variations of like nature. In other words, acquired
habit is probably not a direct determinant of an inherited
variation of like nature in instinctive behaviour. Apart from
such influence the only source of variations which can be
assigned is either the differential division of nuclei in preparation for the process of fertilization,† or the process of fertilization itself. The union of perhaps differentiated germinal
substance from two distinct parents affords the opportunities
for the admixture and compounding of hereditary qualities in
the two samples, from which variations favourable or the
reverse may arise.

It is now generally recognized, however, that the origin of
variations is a problem quite distinct from that of the survival
of those whose direction is favourable to that end. The theory
of natural selection, as such, does not pretend to offer any
explanation of the manner in which variations arise ; though
of course a complete theory of organic evolution must assign
the antecedents and conditions of organic progress in all its
varied phases. We know that variations do occur ; we know,
too, that more individuals are born than survive to procreate
their kind ; and, on the theory of natural selection, we draw
from these data the conclusion that, on the average, the

* *Supra*, p. 36.   † *Supra*, p. 13.

animals that escape elimination are those in which the variations are of such a nature as to conduce to this end.

It will be seen that, on the hypothesis of organic heredity, thus briefly sketched, continuity can, in strictness and, as we may phrase it, in its first intent, only be predicated of the germinal substance ; but that this substance gives rise to products—active vigorous animals behaving in certain ways, each after his kind—which hold similar germinal substance in trust for future use. Natural selection deals with the trustees ; and if they succumb, that which they hold in trust is lost. To put the matter in another way : Nature says to the germinal substance, "By your products you must be judged in accordance with the criterion of utility and efficiency." Practical use in the give and take of active life is the touchstone of all behaviour which makes for survival. This being secured, there may be a balance of behaviour for other purposes. But in animals the balance is not of large amount, and other purposes have not taken form and direction. It should be clearly noticed that, on the hypothesis we are considering, use is the test of survival, and though it is not the direct cause of variations, it affords their sanction in survival. That animal escapes elimination whose behaviour is of practical use ; and it holds in trust for the future a store of germinal substance from which is produced a successor capable of behaving in like manner.

The whole drama of organic evolution may be regarded as the realization in a succession of individuals of the evolving potentiality of continuous lines of germinal substance. The successive individuals die—but the germinal substance lives on in their heirs, if they have any. In virtue of what intimate and hidden structure or disposition of parts the germ possesses this potentiality we do not know. The ovum of a dog is a microscopic speck less than one-hundredth of an inch in diameter ; the sperm is far more minute. They unite, and their nuclei coalesce. The cellular product divides and subdivides. The cell colony absorbs nutriment from the maternal tissues. Division proceeds apace, and the cells are marshalled and ordered in embryological development ; definite tissues are formed ; the stages of their genesis can be predicted with accuracy ; and in due time a puppy is born which shall grow to the likeness of its parents and behave as they behaved. We can trace the succession of events ; we see that they form a related series ; we have good reason for believing that the state of matters at any one moment is the antecedent condition of the state of matters at the succeeding moment. More than this science cannot say. The underlying cause is, for science, hidden in the mists of the unknown. Even for metaphysics it is but part of the force that beats through the universe and makes it not a chaos but a cosmos—a force known to us only in its effects.

It will thus be seen that the conception of continuity in organic evolution has, broadly considered, a threefold aspect. First, there is the continuity of the germinal substance through whose reproductive behaviour under the appropriate conditions embryological development occurs ; secondly, there is continuity in this embryological development, stage by stage, from the fertilized ovum to the adult which is its final product and expression ; thirdly, there is continuity in these final products, in the animals whose organic, instinctive, and intelligent behaviour lie open to our study and investigation. The first is germinal, the second developmental, the third evolutional continuity.

Before attempting to summarize some of the contributions afforded by our inquiry towards the doctrine of continuity in the last of these three aspects, we must pause for a moment to consider how far and in what sense continuity can be predicated of mental development.

We have regarded the conscious situation as the psychical aspect of a nerve-situation in the sensorium ; and the nervous system, capable of behaving in this way, is in developmental continuity with the germinal substance of the fertilized ovum. But what shall we say with regard to the psychical aspect ? Two hypotheses seem open to us, each of which presents difficulties, but of different kinds. The first is, that when the organic development of the nervous system reaches a certain level and order of complexity consciousness emerges, how and whence we know not. The second is, that consciousness is developed from sentience, which is the concomitant of all organic behaviour ; which accompanies life wherever it occurs, and therefore shares the continuity of the germinal substance.

The difficulty inseparable from the first hypothesis, is that it is contrary to the analogy of all that we know or infer elsewhere throughout the realm of nature. Huxley[*] likened its emergence to the production of heat when an iron bar is struck by repeated blows of the hammer. But this analogy will not hold ; for heat is a mode of energy, and only emerges through the transformation of other and pre-existing modes of energy. A certain amount of the energy of motion in the massive hammer-head is transferred to the iron rod, and assumes the form of that molecular vibration which we call heat. And by what amount the one is the gainer, by that amount is the other the loser. But we have no reason to suppose that the like takes place in the origin of the mental concomitants of neural changes. No portion of the brain's store of physical energy is drained off to form the rivulet of consciousness. Now, whenever we speak of a product elsewhere in nature, we mean a specialized bit of something pre-existent. Water is the product of pre-existing oxygen and hydrogen. Heat is the product of other forms of energy. But this is not so on the first hypothesis, according to which consciousness emerges when the functional activity of the nervous system reaches a certain level and order of complexity. The mental concomitants are not "products," in the recognized sense of the term. Furthermore, although on this hypothesis we may still speak of what was termed above evolutional continuity in the mental concomitants, there is nothing analogous to either developmental or germinal continuity.

On the second hypothesis, according to which sentience is the concomitant of all organic behaviour, such developmental

* "Collected Essays," vol. i., p. 239.

and germinal continuity, or their analogues in the psychical order of being, are rendered conceivable. Consciousness is regarded as a developed form of sentience. But the sentience is wholly hypothetical. It is at best a "may be," and its existence is incapable of proof. And science is rightly impatient of hypotheses the validity of which cannot in any way be verified. Our safest course, therefore, is to accept that which is common to both hypotheses, evolutional continuity, and for the rest to be content with a confession of ignorance.

We have already drawn attention to the fact that mere sentience, if it exists, has no power of guidance over organic behaviour ; but consciousness, when it emerges, is a concomitant of nervous processes which determine the nature and direction of such nerve-changes as are the antecedents of intelligent behaviour. The steps by which this control is established are unknown. It is, indeed, probable that conscious guidance arises as an accompaniment of the differentiation of controlling centres from the automatic centres of the nervous system ; but of how this takes place we are as ignorant as we are of many other differentiations in the course of embryological development and evolutional progress. Of those nervous arrangements within the brain which are the physiological concomitants of the far later mental processes of reflection, abstraction, generalization, and the formation of ideals, we are, if it be possible, even yet more profoundly ignorant. Nor would it serve any good purpose to indulge in speculation where there are not even the data to enable us so much as to hazard a probable guess. The utmost we are justified in attempting is to show how organic behaviour leads up to and affords the requisite data for the exercise of intelligence, and how both supply the necessary preliminary stages in the development and evolution of what, following Dr. Stout, we have termed ideational process. This we have endeavoured to do in preceding pages ; and all that is now required is to conclude our inquiry with a brief summary by which the results, as affording some basis for evolutional continuity, may be focussed.

We regard reflex action and instinctive behaviour, broadly considered, as genetically prior to that which is intelligent. Their development in the individual and their evolution in the race are reached by the differentiation and integration of nerve-centres. In the abdominal region of the crayfish, for example, special centres are differentiated for the behaviour of each pair of swimmerets ; but these are so integrated that the whole series of like abdominal appendages swing rhythmically with co-ordinated movements. Now, when a sensorium is developed, it does not have to group by an act of conscious selection and deliberate arrangement the multiplicity of scattered sensory data which it receives ; it does not have to organize from diverse and hitherto unrelated elements some sort of system in experience : it receives them as a physiological heritage already grouped, and to some extent organized. Stimulus and response are organically linked ; and within the response inherited co-ordinations, often exceedingly complex, afford a correlated group of sensory data. Just in so far as organic heredity has provided a working system of bodily parts, does consciousness receive systematic information of their orderly working. No doubt it is true that the development and evolution of the sensorium proceeds *pari passu* with the development and evolution of reflex actions compounded and co-ordinated to give rise to instinctive behaviour. No doubt the progress of the one is in close touch and relation with the progress of the other ; for such relation receives the emphatic sanction of utility. Still it is none the less true that in individual development, as in racial evolution, the organic takes the lead. What is intelligently acquired is something added to that which has been engrained, through natural selection or otherwise, as a potentiality of the germinal substance. What we have first to note, then, is that organic evolution provides ready-grouped data to consciousness.

The second point is, that the germs of abstraction and generalization, or rather processes which are the precursors of abstraction and generalization, arise, and cannot fail to arise, in the genesis of experience from the performance of inherited responses, and from the coalescence of their results into a conscious situation. To a quite young chick I gave pieces of yellow orange peel, which were found to be distasteful and rejected. In Dr. Stout's phraseology, they acquired meaning in experience. Can one doubt that the colour and taste were thus rendered predominant, and that the shape, size, and other qualities of the bits of orange peel remained practically unnoticed ? Shortly afterwards the chick was given chopped and crumbled egg ; the fragments of " white " were eaten, but the bits of hard-boiled yolk were untouched. They possessed a sufficient general resemblance to the orange peel to carry the same meaning. In many ways particular qualities of objects are emphasized in so far as they incite to behaviour ; they form centres of biological interest, just as the abstract quality of ideational thought is the centre of rational interest on a higher plane of mental development. And in many ways objects presenting certain salient features in common, amid differences which remain unnoticed, are unconsciously grouped as the starting-points of similar perceptual situations, just as in the generalization of ideational thought similar relationships are deliberately grouped as the starting-points of like conceptual situations. Both are purposive and have an end, which we as investigators are able to assign ; but only for reflection and conceptual thought are they also purposeful—the end being foreseen and realized, not only by the investigators, but by the agent concerned. And the purpose or end itself is in the two cases different. In the one case it is the biological end of practical behaviour ; in the other case it is the rational end of explanation—abstraction and generalization being deliberately used as a means to this latter end. The question has again and again been asked : Do animals reason ? And different answers are given by those who are substantially in agreement as to the facts and their interpretation, but are not in agreement as to their use of the word "reason." Perhaps, if the question assume the form—Are animals capable of explaining their own acts and the causes of phenomena ?—the position of those who find the evidence of their doing so insufficient may be placed in a clearer light. This is what is generally meant by the statement that animals have probably not reached the level of rational beings.

But even if they have not reached this level, their perceptual processes supply the antecedent conditions which are necessary if this level is to be attained in the course of further evolution. We have seen that, even in relatively simple cases, where conscious situations mark only the beginnings of intelligence, there is a biological emphasis of some, rather than others, among what we call the qualities of objects, and there is a grouping, on biological grounds, of certain things which have some quality in common—such, for example, as being fit for food. Here we have at the outset of perceptual development the germs of processes which are the precursors of the abstraction and generalization of ideational thought. And in the more complex conscious situations of the higher animals these processes attain to such degree of development as is necessary to secure more difficult and more remote biological ends, until all that is necessary, for their rational use, is the quickening touch of a new purpose, that of explanation.

We have seen that, through what Dr. Stout terms "manipulation," and Professor Groos "experimentation"—names applied to a type of behaviour widely exemplified among the higher animals,—things, as the nuclei of conscious situations, become differentiated from the environment. One can hardly question that a fly to the trout, a ball to the kitten, a bone to the puppy are things distinguished from their surroundings, and that they become marked off as special centres of interest. Here on the perceptual plane is a process which is the antecedent of the conception of quasi-independent objects on the ideational plane. For rational thought the thing, as object, is not only the centre of a practical situation leading to behaviour of direct or indirect biological value, but is the nucleus around which we build all the qualities which are ascertained by more elaborate manipulation and experimentation carried out deliberately and of set purpose for rational ends. It becomes capable of definition with the aim of explaining what are its characteristics as an object.

There can be little doubt that the higher animals become intimately and practically acquainted with their environment. The dog who accompanies his master in many a ramble, the horse who carries him again and again over all the surrounding country, has a good perceptual knowledge of a somewhat extended environment. And this, again, is the precursor of the far more extended conceptual knowledge which leads up at last to a rational conception of the universe of objects in their varied relationships. But only through the concentration of thought rendered possible by much true abstraction and generalization,—only through disentangling the relationships and regrouping them for the purpose of framing an ideal scheme,—only, in short, by explanation and for the sake of explanation is this difficult process brought to a more or less successful issue.

Again, there can be little doubt that the higher animals, in the course of experience begotten of behaviour, reach a perceptual sensing of the bodily self, through experience derived from the non-projecting senses, in pain and sickness, and often, we may hope, in the sense of well-being, and the joy of existence. They do not probably set this self in antithesis to the not-self. That comes with reflection, and is the result of ideal construction based on the analysis of experience, with a view to reaching some explanation of the genesis of experience. But in their perceptual awareness of the embodied self, they have that kind of consciousness which affords the necessary data, for the later conception of the self—when experience is polarized into its subjective and objective aspects and thus is explained, so far as science can explain it ; suggesting, indeed, long ere science has attained this end, metaphysical explanations by reference to underlying causes—too often accepted as an easy substitute for the difficult tracing out of the antecedent conditions which science endeavours painfully and by slow steps to formulate.

It is unnecessary to do more than remind the reader that

we have found that such processes as attention and imitation pass through instinctive and intelligent stages which are the precursors of the ideational stage, where they reach a higher expression as deliberately conscious acts. In the young bird that instinctively pecks at some small, perhaps moving, thing, which forms the starting point of a piece of responsive behaviour, we have attention in the germ. When experience has caused the thing to acquire meaning, attention passes into a succeeding intelligent phase ; but only when we desire to explain this meaning, and attention thus has a deliberate purpose, do we find it entering upon its higher ideational career. So, too, as we have seen, imitation is at first a specialized form of instinctive behaviour, where the response is seen to resemble that which stimulates it. Later it becomes intelligent when the repetition of the imitative behaviour is due to the satisfaction it introduces into the conscious situation. Then, at last, it reaches the ideational stage, where reflection gives rise to an ideal, which is to be realized in conduct. The imitation by the child of its older companions is at first probably intelligent ; but when the child begins to consider why it imitates these and not those among its companions, he is passing to the ideal stage, and imitation becomes the sincerest form of hero-worship. The boy who merely imitates his elder brothers playing at soldiers because he gets satisfaction from so doing, becomes the subaltern who has his ideal soldier, and will face death firmly rather than fall below his conception of how such a soldier should behave.

We need not again attempt to indicate how among animals we have the perceptual precursors of the æsthetic and ethical concepts. But we may remind the reader that we endeavoured to show that intercommunication had its foundation in instinctive sounds ; and that it passed into the intelligent stage in the perceptual life, when these sounds acquired meaning, and hence became guides to behaviour. This is especially instructive from our present standpoint, since it is probable that the passage of communication from the indicating to the descriptive stage afforded the conditions under which rational thought was evolved. For such thought it is essential that attention should be focussed on the relationships of things. And no description is possible without making distinctly present to consciousness these relationships, in time and space, the data for which are abundantly present in the perceptual life, though lurking in the background, and needing something to fix them and to aid consciousness in distinguishing them clearly. In descriptive communication parts of speech, or their initial equivalents, afford fixation points for these relationships, and serve to render them distinct. If the reader will try to describe even the simplest occurrence without introducing the symbols for the relations which the events bear to each other, his failure will serve to bring home how essential a feature this is. In social communication, then, we probably have the key to the passage from perceptual to ideational process; and in this passage description is the antecedent of, and affords the conditions to, explanation. Words, moreover, as we have already said, form the pegs upon which we can hang up, for ready reference, the products of abstraction and generalization, or, to modify the analogy, they form the bodies of which these products are the rational soul.

If we are ever to trace the passage from the instinctive through the indicating stage of communication, and so onwards through the beginnings of description to its higher levels, and thus to the use of language as a medium of explanation, it must be through child-study. In every normal human child the passage does actually take place, though, no doubt, in a condensed and abbreviated form as an epitomized recapitulation in individual development, of the steps of evolutional progress. Thus we may obtain a key to the solution of one of the most difficult problems in evolution by continuous process—that of the transition from animal behaviour to human conduct.

INDEX

A

Abstract and general ideas, 57
Abstraction, 166; germs of, 332 ff.
Acceleration, 250
Accommodation defined, 36
Acquired characters, inheritance of, 35, 110
Acquired instincts (Wundt), 66, 106.
Acquisition defined, 36; ultimately dependent on natural selection, 289
Adaptation defined, 37
ADDISON on instinct, 63
Æsthetics, animal, 270
Afferent and efferent impulses, 32, 101
Aid, mutual, among animals, 227
Ammophila mode of stinging prey, 75; of carrying prey, 76; deposition of egg, 77; intelligent behaviour of, 127
Amœba, 296
Antlers of deer, 15
Ants, behaviour of, 123; inter-communication of, 198; social communities of, 205
Aporus, intelligent behaviour of, 126
Appreciation, germs of, 273
Ardour of male in courtship, 269
Argyromœba, instincts of, 79
Arrest of development in egg, 14
Association in coalescent situation, 46
Attention, 242
AUDUBON on American night-hawks, 261

B

AVEBURY, Lord, on ants, 198; on Van, 200; on aphides and ants, 214; on slave ants, 215; on intelligence of ants, 218

B

BALDWIN, Prof. Mark, on organic selection, 37 (note), 115; on functional selection, 163; on imitation, 179 ff.; on projective stage of development, 275
Batesian mimicry, 164
BECKSTEIN on canaries, 262
Bees, homing of, 131; social communities of, 205
Beetle soliciting food from ant, 213
Bembex mode of carrying prey, 76
BETHE, Dr., on instinctive behaviour of ants, 217
BINET, M., on infusoria, 6
Biological value of play, 250; purpose, 294; aspect of animal behaviour, 305
Birch-weevil, leafcase of, 121
Birds, instinct of, 84
Bison, behaviour of the, 226
BLACKBURN, Mrs. Hugh, on instinct of cuckoo, 90
BLOCKMANN, Dr., on *componotus*, 210
BOLTON on goldfinches' nests, 136
Bower-bird, observations on, 261, 273
BUCKMAN, Mr. S. S., on speech of children, 203
BUDGETT, Mr. John S., on nest-building, 135
Bullfinch, nest of, 135

C

CAMERON, Mr., on mimetic insects in ants' nest, 212
Canaries' nest, building of, 135
Canon of interpretation, 270
Capacity, innate, 176
Capuchin monkey, imitation in, 188, 278
CARPENTER, W. B., on water-beetle, 299
Catasetum, fertilization of, 29
Cats, Prof. Thorndike's experiments on, 147, 184
Causation, idea of, 257
Cell-division in egg, 14
Cerceris, instincts of, 74; locality studies of, 129
Chalicodoma, parasities of, 78; Fabre's observations on, 130
Chick swimming, 85; instincts of, 85 ff.; imitation in, 183
Child-study, desirability of, 155, 337
Choice, apparent in Paramecia, 9; in the pairing situation, 266
Ciliary action in Paramecium, 4-10
Circular process (Baldwin), 181
Clepsine, behaviour of, 159
Coalescence in conscious situation, 46
Coincident variations defined, 37; survival of, 115, 174
Communities, social, of bees and ants, 205
Companion as centre of special interest, 244
Componotus, communities of, 210
Conation and impulse, 187, 235
Concept, nature of, 166
Condensation of experience, 162
Conduct implies motive, 60; and ideal, 278
Congenital responses, 41
Conjugation in Paramecium, 4
Connate instincts, 66, 69
Conscience, ambiguity of word, 281
Conscious accompaniments of certain organic changes, 42; aspect of instructive behaviour, 99
Consciousness, as accompaniment and as guide, 34; effective, de-

fined, 43; as heir to organic estate, 52; as epiphenomenon, 306
Consentience, 53, 62
Consonance of biological and psychological end, 286, 316
Constancy of environment leads to stereotyped behaviour, 172
Continuity in evolution, 324; three-fold aspect of, 329
Control, the sign of effective consciousness, 43
Co-ordinated acts, 69, 100; inherited, 94, 95
Corporate behaviour, 14
Courtship in animals, 259
Coyness of female birds, 264 ff.
Cranial sense-organs, 301
Crayfish, reflex action in, 298
Creation, special, 297
Criteria of effective consciousness, 43; of intelligence, 120
Cruelty in cat, 277
Cuckoo, instinct of nestling, 90

D

DARWIN, Charles, fig. of sun-dew leaf, 26; of Venus's fly-trap, 27; of catasetum, 31; on earthworms, 157; on social life of animals, 225; on human ancestry, 229; on play and practise, 259; on sexual selection, 262 ff.; on law of battle, 313
DAVIS, Prof. Ainsworth, on limpets, 155
DEAN, Dr. Bashford, on chick swimming, 85
Deceit in animals, 280
Deferred instincts, 70
Definiteness of instinctive behaviour, 66
Description, involves relational terms, 202
Didunculus, changed habits in, 221
Differentiation and integration of nerve centres, 167
Disintegration of instincts, 175
Diving, instinctive, 86
Dog, observations on, intelligence of, 141 ff., 152, 200, 271, 322

Duckling, inherited co-ordination in, 96
Dytiscus, instinct of, 104

E

Earthworms, Darwin's observations on, 157 ff.
Education in play, 255, 320 ff.
Effective consciousness defined, 43
Efferent and afferent impulses, 32, 101
Egg, cell-division in, 14
EIMER, Th., on instincts of solitary wasps, 73; on origin of instincts, 107
Emotions, and feelings, 235 ff.; psychological nature of, 246; evolution of, 282
Energy stored in cell, 23
Equilibrium, tendency to, 296
Eris'alis, mimicry of, 164
ESPINAS, Prof., on social life of animals, 230
Ethics, animal, 270
Evolution of organic behaviour, 35; of consciousness, 61; of instinctive behaviour, 106; of intelligent behaviour, 155; of social behaviour, 225; of feeling and emotion, 282; of animal behaviour, 295; as continuous, 324 ff.
Experience, of value for future guidance, 41; is it inherited? 48, 97; condensation of, 162
Experimentation, 251, 253
Explanation, characteristic of later phases of mental development, 58, 257
Explosive nature of cell, 21
Expression of emotions, 247
External stimuli to instinctive behaviour, 102

F

FABRE on behaviour of *Sphex*, 77, 171; of *Chalicodoma*, 78, 129; of *Leucopsis*, 79; of *Pompilus*, 129
Faculty, instinctive, 64
Falcons, training of, 137

Fear in birds not inherited in specific direction, 49, 110
Feelings and emotion, 235 ff.; evolution of, 282; feeling-tone, 240
Ferns, fertilization of, 24
Fertilization of ferns, 24; of *Valisneria*, 28; of orchids, 29
FINN, Mr. Frank, on the acquisition of experience by young birds, 50
Fission, reproduction of Paramecium by, 4
Flight, instinctive, 86
FOREL on *Componotus*, 211
FOSTER, Sir Michael, on consciousness accompanying reflex action in pithed frog, 33
Frog, reflex action in, 33, 299, 300
Functional selection, 163
Fungus garden of ants, 216

G

GARNER, Mr. R. L., "The Speech of Monkeys," 198
Gas-engine, analogy of, 20
General and abstract ideas, 57; generalization, 166; germs of, 332 ff.
Generic image, 162; situations, 163
Germinal substance, continuity of, 328
GOULD, Dr., on humming-birds, 273
GREEN, Mr. E. G., on ants, 210
Greenfinch, nest of, 135
GROOS, Prof., on instinct, 64; origin of, 116; on imitation, 187; on animal play, 248 ff.; on "Love Play," 259; on coyness of female birds, 264; on choice in mating, 267; on make-believe, 280

H

Habits and habitual acts, 107, 177
HAGUE on ants, 199
HAMERTON, P. G., on trained dog, 152
HANCOCK, Dr. John, on cuckoo, 92
Heredity and circumstance, 39; twofold aspect of, 40; relation of to use, 170, 176; in evolution as continuous, 326

Homing of bees, 131
Homœopathic influence defined, 36
Honey-pot ant, 215
House-martin, nest-building of, 112
HUDSON, Mr. W. H., on fear in birds, 49, 50, 110; on animal gladness, 317
Hunting play, 254
HUXLEY, T. H., on reflex action in frog, 300; on consciousness as epiphenomenon, 306 ff.; on consciousness as product of nervous changes, 330
HYDE, Mr., on king crab, 298
Hydractinia, colonial polype, 206

I

Ideals, distinguish ethics, 278
Ideational stage of mental development, 59
Imitation, 179 ff.; three stages of in child, 192
Impulse in intelligent behaviour, 60; Prof. Thorndike's use of the term, 186; connection of with conative process, 235
Independence of automatic and controlling centres, 43
Infant, congenital responses in, 54
Influence of intelligence on instinct, 168
Inheritance of acquisitiveness, 40
Innate capacity, 176; likes and dislikes, 119
Insects, instinctive behaviour in, 71; intelligent behaviour in, 123
Instinct, broader and narrower view of, 99; primary and secondary, 107, 109; influence of, intelligence on, 168; priority of, to intelligence, 173; disintegration of, 175
Instinctive behaviour defined, 63; in insects, 71; in birds, 84; conscious aspect of, 98
Integration and differentiation of nerve centres, 167
Intelligence lapsed, 107; influence of, on instinct, 168; of ants, Lord Avebury on, 218; biological importance of, 310
Intelligent behaviour, 117; evolution of, 155
Intelligent process distinguished from rational, 59, 138
Intercommunication, 193, 336
Interest, 243
Internal factors in instinctive behaviour, 102
Irritability, fundamental, property of protoplasm, 240, 296

J

JAMES, Prof. Wm., theory of emotions, 246, 292
Jays, bathing of, 89; mode of taking food, 94
JENNER on cuckoo, 92
JENNINGS, Dr. H. S., on behaviour of Paramecia, 5

K

KERNER, Dr., on sun-dew, 26; on sensitive Oxalis, 27; on Valisneria, 29
King-crab, reflex action in, 298
KNIGHT, Andrew, on Norwegian ponies, 109
KROPOTKINE, Prince, on mutual aid among animals, 227

L

Lamarckian hypothesis, 169, 170, 177
Language, nature of, 195
LANKESTER, Prof. E. Ray, on small-brained mammal, 168
Lapwing, instinctive behaviour of, 113
Law of battle, 313
Leech, observation on, 159
Leucopsis, instincts of, 79
LEWES, G. H., on lapsed intelligence, 107
Limpets, observations on, 155
LINDLEY, Dr., on children, 141
Locality, studies by wasps, 128
LOCKE, John, limitations of animals, 166

M

MACKENZIE, Prof. J. S., on ethics, 278; on conscience, 281; on ambiguity of word " pleasure," 285

Make-believe, 280
Mammals, early small-brained, 168
Manipulation (Stout), 251
MARCUAL, Prof., on instincts of *Cerceris*, 74
MARSHALL, Mr. H. R., on instinct, 66
Martin, nest-building of, 112
MARTINEAU, James, on pleasure and pain, 284
MAUPAS, M., observations on, infusoria, 4
MAYER, Dr. A. G., on mating, instinct of moths, 83
McCook, Dr., on ants, 214
Meaning, Dr. Stout's use of term, 46, 243, 268
MEDLICOTT, Mr. H. B., on behaviour of wild pigs, 196
Megapodes, instinctive flight of, 87
Meloë, instincts of, 81
Mental development, stages of, 48, 56
MERCIER, Dr. Charles, on criteria of intelligence, 120
Metaphysical explanations, 19; aspect of instinct, 64; of impulse, 237; of purpose, 294; of will, 307; of life, 323
MILLS, Prof. Wesley, on social influence on puppy, 220
Miltogramma, parasitic fly, 134
Mimicry, Batesian and Müllerian, 164; intelligent aspect of, 311
Modifiability, 171
Modification, defined, 36; relation of, to hereditary characters, 169
MÖLLER, Herr, on fungus garden of ants, 216
Monistic hypothesis, 309, 315
Monkey, capuchin, imitation in, 188
Monodontomerus, instincts of, 79
Moor-hen, diving of, 89
Moths mating, instinct of, 83
Motive in rational conduct, 60
Movement plays, 251
MÜLLER, Prof. Max, on barrier between brute and man, 204
Müllerian mimicry, 164, 165
Mystery of life, 18

N

Natural selection, shielding of chicks from, 111; under uniform and variable circumstances, 175; in playtime of life, 319
Nervous arc, 33; system of higher animals, 297
Nest-building, observations on, 135
NOIRÉ on concept, 166
Norwegian ponies, 109
Nucleus division, 12

O

Object and subject, 245, 276
Octopus, intelligence of, 157
Œcophylla, behaviour of, 210
Orchids, fertilization of, 29
Organic basis of differentiation of consciousness, 53
Organic behaviour in development, 15
Organic selection, 37 (note), 115
Overproduction of movements, 163
Oxalis, sensitive, behaviour of, 27
Oxybelus, mode of carrying prey, 76

P

PALEY, definition of instinct, 64
Paramecium, behaviour of, 3, 296
Partridge, note of young, 93
PECKHAM, Dr. G. W., on instinct, 65; on solitary wasps, 72 ff., 126 ff.
Pecking instinct of chicks, 93
Peewit, note of young, 93
Pelopæus, instincts of, 72
Perceptual stage of mental development, 59
Personality, 245, 257
Pheasants, note of young, 92; inherited co-ordination in, 95; plumage of Argus, 262
Philanthus, prey of, 73; mode of stinging prey, 74
Physiological aspect of animal behaviour, 295
Pigeons, nests of, 136
Pigs, wild, behaviour of, 196
Plants, behaviour of, 24

Plastic period of life, 167
Plasticity of tissues, 40; of behaviour, 172
Play of animals, 248 ff.; biological value of, 250; psychological aspect of, 256, 311, 316
PLAYNE, Mr. H. C., on pigeons' nests, 136
Pleasure, 241; ambiguity in word, 285
Polistes, locality studies of, 131
Pompilus, mode of carrying prey, 76; Fabre's observation on, 129
Presentative elements distinguished from re-presentative, 46
Primary instincts (Romanes), 107
Projective stage of mental development, 275; senses, 304
Pronuba, instinct of, 82
Propensity, instincts as, 64; congenital, 177
Protoplasm, fundamental properties of, 296
Psychological aspect of play, 256; purpose, 294; aspect of animal behaviour, 315

R

Rational process distinguished from intelligent, 59, 138
Reflex action, 31, 35, 298 ff.; relation of instinct to, 70
Relationships, importance of, 202
Re-presentative elements distinguished from presentative, 46
Rhynchites, instinct of, 121
ROMANES, G. J., on "discrimination" and "perception" in plants, 32; on instincts of solitary wasps, 73; definition of instinct, 99; on primary and secondary instincts, 107, 109; on ants, 126; on general ideas, 166; on animal communication, 201; on cruelty in cat, 277
ROMANES, Miss, observations on capuchin monkey, 188, 278
Roots of spinal nerves, 299
ROTHNEY, Mr. G. A. G., on Indian ants, 212

S

SCHNEIDER on octopus, 157
SCOTT, Dr. D. H., on fern fertilization, 25
Scratching in duckling, 96
Sea anemone, diffused nervous system of, 32
Secondary instincts (Romanes), 107, 109
Segmental nature of central nervous system, 299
Selection, functional, 163; natural, shielding of chicks from, 111; under uniform and variable circumstances, 175; in playtime of life, 319; sexual, 261 ff., 313
Self, as ideal construction, 239
Sentience, 62; origin of, 330
Sexual selection, 261 ff., 313
SHARP, Dr. D., on birch-weevil, 121; on *Œcophylla*, 210
SHELLARD, Mr. E. J., observations on staghound, 144
SHERRINGTON, Prof., on emotion, 292; on spinal animal, 298 ff.
Shock effects of physiological, 302
SIMCOX, Miss Edith, quoted, 320
Sitaris, instincts of, 82
Slave ants, 215
Snails, observations on, 157
Social behaviour, 179; evolution of, 225
Solitary wasps, instincts of, 72 ff.; intelligence of, 126 ff.
Solomon Islands, rats of, 222
Sounds emitted by young birds, 92
SPALDING, Douglas, on newly hatched turkeys, 49; on instinct, 99
Special creation, 297
Speech, connection of, with rational process, 58, 233; so-called, of monkeys, 198; of children, 203; aids passage from perceptual to ideational process, 337
SPENCE on instinct, 63
SPENCER, Mr. Herbert, on instinct and reflex action, 70; on play due to surplus vigour, 248; on pleasure and pain, 284, 287; on survival, 288

Sphex, mode of carrying prey to nest, 77, 171
Spiders placed in crotch by wasps, 133
Stereotyped behaviour, 172
Sticklebacks, observations on, 130
STOUT, Dr. G. F., on "meaning" for consciousness, 46, 243, 268; on ideational and perceptual stages of mental development, 59, 271; on octopus, 157; on conative process, 235; on the self, 239; on manipulation, 251; on emotion, 293
STRANGE, Mr., on bower-bird, 261
Subject and object, 245, 276
Sun-dew leaf, behaviour of, 25
Swimming, instinctive, 85

T

TAIT, Lawson, on begging-cat, 109
TARDE, M., on imitation as a social factor, 179
Tendencies, congenital, 177
THOMAS, Mr. Oldfield, on rats of Solomon Islands, 222
THORNDIKE, Dr., on swimming of chick, 85; experiments on intelligence, 147 ff.; experiments on imitation, 179, 183 ff.
Tradition, animal, 220
Trial and error, method of, 139

U

Unity of perceptual process, 240; of biological purpose, 297
Use, super-normal, 169; relation of, to heredity, 170, 176

V

Valisneria, fertilization of, 28
Variation defined, 36; origin of, 327
Venus's fly-trap, behaviour of, 26
Vigour, play due to surplus, 248
Volition as conative, 238

W

WALLACE, Dr. A. R., on sexual selection, 264
WALLASCHEK, Mr., on play as surplus vigour, 248
Wapiti, antlers of, 16
WASMANN, Dr., on insects associated with ants, 213
Wasps, solitary, instincts of, 72 ff.; intelligence of, 126 ff.
WEIR, Mr. Jenner, on canaries, 135
WEISMANN, Prof., on origin of instinct, 108
WHITMAN, Prof., on *Clepsine*, 159
WHITMEE, Rev. S. J., on tooth-billed pigeon, 221
Will, metaphysics of, 307
WILLISTON, Dr. S. W., observation on *Ammophila*, 127
WOOD, Mr. Foster, on hen-swimming, 86
WORCESTER, Dr., on megapode, 87
WUNDT, Prof., on instinct, 65, 99, 106

Y

Youth, plasticity of, 167
Yucca moth, instincts of, 82

THE END

Made in the USA
Monee, IL
07 July 2026

56646384R00204